SOLAR ECLIPSES
2024–2027
WHERE AND WHEN TO EXPERIENCE TOTALITY

BONUS ANNULAR ECLIPSES
2023 & 2028 INCLUDED

SHERIDAN WI

www.bradtg

T0016983

Bradt Guides Ltd,
The Globe Pequot Press Inc, USA

Bradt GUIDES
TRAVEL TAKEN SERIOUSLY

AUTHOR

Sheridan Williams is a Fellow of the Royal Astronomical Society and a leading 'eclipse chaser'. In 1966 he built his own telescope and has since travelled to see 18 total and two annular solar eclipses. Publications include a book on *UK Solar Eclipses* and five Bradt guides covering eclipses from 2001, plus numerous contributions to magazines. He has appeared on television and radio, including presenting Sky TV's total eclipse programme from Cornwall in 1999. He writes regular features for his local newspaper and lectures widely to astronomical societies and other interest groups. He is also an astronomy travel consultant and guide, leading groups to see various astronomical events.

First published March 2023
Bradt Guides Ltd
31a High Street, Chesham, Buckinghamshire, HP5 1BW, England
www.bradtguides.com
Print edition published in the USA by The Globe Pequot Press Inc,
PO Box 480, Guilford, Connecticut 06437-0480

Text copyright © 2023 Bradt Guides Ltd
Maps copyright © 2023 Bradt Guides Ltd; includes map data © OpenStreetMap contributors
Photographs copyright © 2023 Individual photographers (see below)
Project Manager: Susannah Lord
Cover research: Sheridan Williams FRAS

ISBN: 9781804690857

British Library Cataloguing in Publication Data
A catalogue record for this book is available from the British Library

Photographs © individual photographers credited beside images and also those from picture libraries credited as follows: Shutterstock.com (S); Superstock (SS).
Front cover Total solar eclipse of 29 March 2006, Libya (Sheridan Williams)
Back cover Partial solar eclipse seen at the Lampam lakeshore, Phatthalung province, Thailand (Wipark Kulnirandorn/S); mid-eclipse, seen from Anaa Atoll, French Polynesia, 11 July 2010 (Howard Brown-Greaves)
Title page, clockwise from top left Total solar eclipse, China 2008 (Nick James); partial solar eclipse observed in the morning at the Western Wall, Jerusalem (John Theodor/S); prominences seen during the 2005 annular eclipse, Moraira, Spain (Sheridan Williams)

Maps David McCutcheon FBCart.S

Typeset by D & N Publishing, Baydon, Wiltshire and Ian Spick, Bradt Guides
Production managed by Zenith Media; printed in the UK
Digital conversion by www.dataworks.co.in

Other credits All lunar limb profiles (LLPs) and sky charts have been produced by Xavier Jubier. Many of the eclipse maps and data are courtesy of Fred Espenak, Jay Anderson and Xavier Jubier. Climate notes are by Jay Anderson and Sean Clarke.

Contents

LIST OF MAPS

FEEDBACK REQUEST

At Bradt Guides we're aware that guidebooks start to go out of date on the day they're published – and that you, our readers, are out there in the field doing research of your own. You'll find out before us when a fine new family-run hotel opens or a favourite restaurant changes hands and goes downhill. So why not tell us about your experiences? Contact us on ☏ 01753 893444 or e info@bradtguides.com. We will forward emails to the author who may post updates on the Bradt website at w bradtguides.com/updates. Alternatively, you can add a review of the book to Amazon, or share your adventures with us on Facebook, Twitter or Instagram (@BradtGuides).

LOCAL CIRCUMSTANCES The timings of each eclipse differ depending on your location, together with the eclipse duration and altitude of the Sun. Details for a variety of locations are provided for each eclipse.

ECLIPSE TRACKS Charts showing the eclipse track are included in the *Where it goes* section for each eclipse described in this guide. A website link to a more detailed map (courtesy of Xavier Jubier; w xjubier.free.fr) is also given – once the relevant web page is loaded, click anywhere on the map to see local circumstances.

LUNAR LIMB PROFILES (LLPs) Diamond ring positions are accurate for those observing near the centre line; see page 17 for more information on libration and the LLP.

SKY CHARTS The sky charts provided show what to look out for during totality.

ECLIPSE FACTS Each eclipse is unique. We have listed facts that are relevant to each eclipse, such as:

- Time of greatest eclipse
- Maximum duration of totality
- Location of greatest eclipse
- Altitude of Sun at greatest eclipse
- Maximum width of path of totality
- Saros Series
- Magnitude
- Obscuration
- Gamma

ECLIPSE DESTINATIONS A–Z In the final chapter of this guide, useful visitor information has been collated on each of the eclipse destinations. See from page 103.

Introduction

There is nothing in Nature to rival the glory of a total eclipse of the Sun. No written description, no photograph, can do it justice.

Sir Patrick Moore CBE, veteran British astronomer, 1999

A total solar eclipse is the most moving experience on Earth for many who witness one. Some feel unrestrained joy at the sight of totality, others a powerful sense of desolation, like this observer in England in 1927:

It was as if the whole Earth were smitten with a mortal sickness… It was all inexpressibly sad and utterly desolate, and when the small golden crescent appeared again behind the Moon, I involuntarily uttered the words 'Thank God for the Sun.'

No-one can really explain why a total eclipse of the Sun has the power to unleash such varied emotions, leaving us so momentously disorientated. After all, the movements of the Earth, Moon and Sun have been predictable for thousands of years – and explicable for hundreds. They are hardly a surprise anymore. Yet they still overwhelm us. I believe it is because of the instinctive realisation, as the eclipse unfolds, that nothing can prevent it. We already know this in a rational way, of course, but the eclipse drives it into our understanding with unequalled drama. Another observer in 1927 wrote:

It is well that we should now and then be made to realise the might and majesty of the universe in which we are the greatest and the least.

As long as eclipses have been predictable, people have been travelling to see them. There are recorded reports well over 200 years old (and numerous reports of solar eclipses thousands of years ago before eclipses could be predicted) – and numbers are increasing as word spreads about this natural wonder. Wherever you go, you will be joining a distinguished club of 'eclipse chasers' (often known as umbraphiles).

◀ *Composite of 646 images showing the full extent of the solar corona, and numerous stars to magnitude 10. Taken from Cerro Tololo, Chile (2019).*

1

Total Solar Eclipses

Every time the Moon passes between the Earth and the Sun, casting its shadow on our planet, the experience for the humans who watch it is different from the last. Every solar eclipse is governed by a multitude of rhythms and subtleties that guarantee its individuality. The Sun's activity, its flares and prominences, the exact positions of the Moon and the Earth – all these factors conspire with numerous others to ensure that each eclipse is unique. Similarly, the aesthetics of an eclipse are influenced by the time of day, the region of the world, the weather and, to a surprising extent, the surrounding crowds or emptiness.

Yet, despite this capacity for infinite variation, eclipses are governed by the enduring laws of celestial mechanics that guarantee us totalities for hundreds of millions of years to come.

EXPLANATIONS

WHY ECLIPSES HAPPEN A total solar eclipse occurs when the Moon moves between the Sun and the Earth, fully blocking our view of the Sun.

This happens because of a heavenly coincidence: the Sun and the Moon, when viewed from the Earth, appear to be the same size. The Sun may seem at first to be the larger but this is because it is so bright. If you stick your thumb and forefinger out to gauge their apparent (or angular) sizes you will find they are the same.

In reality, the Sun's diameter is 400 times that of the Moon. But, with perfect compensation, it sits 400 times further away. This symmetry has not always existed, but scientists have calculated that things will remain this way for another 650 million years, after which the Moon will drift too far from the Earth for it ever again to be able to cover the Sun completely, resulting in annular eclipses only.

During the year the Earth takes to complete its journey around the Sun, the Moon speeds around the Earth about 13 times. Once in each of these cycles (known as synodic months) the Moon passes between the Sun and us and we see a New Moon. We do not see a monthly solar eclipse, though, because the Moon's orbit is at a slight angle to that of the Earth – so the Sun and Moon 'miss' each other as they pass across our skies.

To understand the orbits relevant to an eclipse it is easier to consider a model in which the solar system is inverted, with the Earth at its centre. This model, still in use for some purposes today, represents not the reality of planetary orbits but what we actually see when we look at the skies. In this model, the Sun's apparent path across the sky is known as the ecliptic.

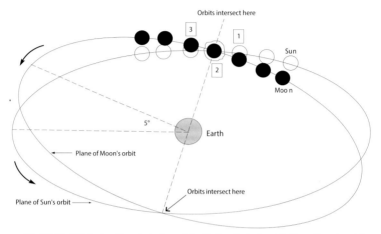

The Moon's orbit is at a 5° angle to the ecliptic (Sun's path in the sky). At positions 1 and 3 there is a partial solar eclipse. At position 2 there is a central eclipse.

As viewed from the Earth, the Moon circles us every 27.5 days, and the eclipse year of 346.62 days is the time for the Sun (as seen from the Earth) to complete one revolution with respect to the same lunar node. The Moon catches up with the Sun every 29.5 days and would pass between it and Earth, causing a solar eclipse if it weren't for those inclined orbits (see diagram, above). But the two orbital paths do, of course, intersect at two points. These points – or nodes – represent two periods each year during which the Moon can indeed pass in front of the Sun. When the Moon passes exactly in front of the Sun, we see either a total or annular eclipse; these are called 'central' eclipses because the Moon passes centrally across the Sun.

If the Sun, creeping along its yearly path, happens to be almost exactly at one of the two intersections when the Moon is racing round on its monthly cycle, there will be a central solar eclipse. If the Sun is a little further from the intersection, there is still the chance that the Moon will at least overlap with it, creating a partial solar eclipse.

In fact, the Sun moves so slowly that it is impossible for it to get past either intersection without the Moon catching up with it at least once, and therefore guaranteeing a solar eclipse of some sort. Sometimes the Moon manages to travel all the way round and catch up with the Sun before it has fully left the intersection – causing two partial eclipses a month apart. Whatever happens during these eclipse 'seasons', there will always be at least two solar eclipses of some sort each year.

Those are the basics behind a solar eclipse. In practice, however, there are various distortions and complexities that give each eclipse its signature. For example, both the Sun and the Moon appear to follow elliptical paths around the Earth rather than circular ones, with the result that their distances from our planet vary. If a solar eclipse occurs when the Moon is at its furthest from the Earth and the Sun at its closest, a rim of Sun remains around the black disc of the Moon – and we see an annular eclipse.

SHADOWS Before we can discuss the kinds of eclipse and how they happen, we need to establish the types of shadow a body can cast. Bear in mind that we are talking about three nearly spherical bodies, Sun – Moon – Earth, and in solar eclipses the bodies are in this order. (For lunar eclipses the position is Sun – Earth – Moon.)

When the Moon is between us and the Sun, it must (by definition) be a New Moon and hence invisible as the Sun illuminates the side of the Moon facing away from the Earth.

The Earth and Moon are both in elliptical orbits

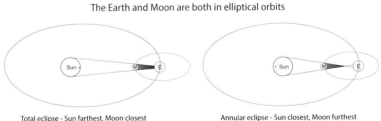

Total eclipse - Sun farthest, Moon closest   Annular eclipse - Sun closest, Moon furthest

TOTAL ECLIPSE The Moon's umbral shadow reaches as far as the Earth. The umbral shadow will be tiny compared to the size of the Earth, and averages 150km in diameter. Outside the umbra a partial eclipse is visible over typically one-third of the Earth's daylit surface.

Total eclipse

What you see

ANNULAR ECLIPSE Here the umbral shadow does not reach the Earth's surface, and the Moon is too small to obscure the Sun completely. Locations not immediately behind the umbra will observe a partial eclipse covering one-third of the Earth's daylit surface.

Annular eclipse

What you see

PARTIAL ECLIPSE Here the Moon's umbra misses the Earth completely, and all you see is the Sun partially covered by the Moon. A partial eclipse is visible over a wide area, typically one-third of the daylit surface of the Earth.

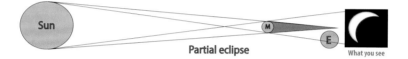

Partial eclipse

What you see

ANNULAR-TOTAL (OR HYBRID) ECLIPSES These can happen on rare occasions when the Moon's umbra is not long enough to touch the Earth's surface at the beginning A (and end C) of an eclipse, but is just long enough to strike the surface B in the middle of the eclipse.

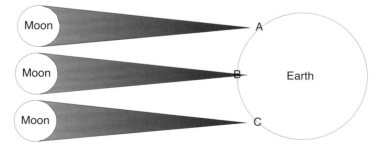

THE SAROS The periodicity and recurrence of eclipses is governed by the Saros cycle, a period of approximately 6,585.3 days (18 years 11 days 8 hours). This cycle was discovered by the Chaldeans possibly as early as 400BC and has been used over subsequent millennia to predict eclipses. It is also thought that the Antikythera mechanism found in the Mediterranean

DIAMETERS
Sun 1,392,530km
Moon 3,476km
Earth 12,756km

DISTANCE FROM THE EARTH

	Min.	Average	Max.
Sun	147,000,000km	149,600,000km	152,000,000km
Moon	356,400km	384,400km	406,700km

The Sun is closest to Earth on 2–4 January, and furthest on 5–6 July.

APPARENT SIZES (in minutes of arc)

	Min.	Average	Max.
Sun	31.47'	32.00'	32.53'
Moon	29.40'	31.46'	33.53'

was used to predict astronomical positions and eclipses. When two eclipses are separated by a period of one Saros, they share a very similar geometry. The eclipses occur at the same node with the Moon at nearly the same distance from Earth and at the same time of year, thus, the Saros is useful for organising eclipses into families or series. Each series typically lasts 12 to 13 centuries and contains 70 or more eclipses.

An eclipse casts a shadow in an arc across the Earth. For that shadow to be identical to the shadow cast by another eclipse, the relative orientations of the Earth, Moon and Sun would have to be exactly the same. This can occur only on the rare occasions when the Moon's position in its orbit around the Earth and the Earth's position in its orbit around the Sun are repeated. This does not happen every time. For example, when one eclipse year has passed and the Earth begins its orbit around the Sun again, the Moon is three-quarters of the way through a cycle round Earth; so it is not back in the same position as it was when the Earth last began an eclipse year. But after the passage of approximately 18 years 11.32 days the Moon is just coming to the end of its monthly cycle when the Earth is coming to the end of its yearly cycle – and they pass through positions very similar to those of 18 years previously. Thus, when an eclipse occurs over a certain region of the Earth, we can predict that, 18 years later, an eclipse will occur at a similar latitude. Since this synchronisation is not perfect, the next

Total Solar Eclipses   EXPLANATIONS

1

NUMBER OF ECLIPSES IN ONE YEAR The maximum number of solar eclipses in any one year is five; the minimum is two. The maximum number of lunar eclipses in any one year is four; the minimum is two. However, there can never be more than seven eclipses in any year, but there must be at least four.

LONGEST POSSIBLE
Total: 7m 31s (example: 7m 29s on 16 July 2186)
Annular: 12m 30s (example: 12m 09s on 14 December 1955)

MAXIMUM SHADOW WIDTH
Total: 269km width of umbral cone
Annular: 370km width of antumbral cone

The width of the shadow when projected on the Earth's surface can be far wider. For example, in 2600 the shadow projected on the Earth is 627km.

The typical speed of the Moon's shadow at the longest eclipse is 1,730km/h.

Scale diagram of Moon-Sun sizes

Most annular
Sun 10.6% larger

Most total
Moon 6.5% larger

FREQUENCY OF ECLIPSES There are approximately 450 eclipses per century made up as follows:

	Solar	Lunar
No. of eclipses	224	226
Total	70	80
Annular	72	–
Annular-total	6	–
Partial	76	146 (lunar partial umbral or penumbral)

OBSCURATION AND MAGNITUDE The fraction of the Sun's area covered by the Moon is known as Obscuration. Magnitude is how much of the solar diameter is covered by the Moon – it has a value between 0 and 1, where 1 is totally eclipsed.

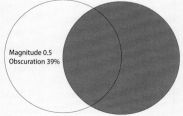

Magnitude 0.5
Obscuration 39%

GAMMA Gamma is measured in Earth equatorial radii and is the distance by which the Moon's shadow axis misses the centre of the Earth.

Gamma = 0 means the Moon's shadow axis passes directly through the centre of the Earth. A positive gamma means that it passes north of the equator; a negative gamma passes to the south of the equator.

Gamma = ± 0.997 means the shadow axis is exactly tangential to the northern (or southernmost) point of the Earth, as the polar radius is less than the equatorial radius. Gamma = ± 0.997 to ± 1.026 means that the shadow axis misses the Earth; however, the eclipse can still be total or annular even though it has no northern (or southern) limit. This type of eclipse is called a non-central total (or annular) eclipse.

NON-CENTRAL ECLIPSES A non-central eclipse is one where the Moon's shadow axis does not touch the Earth's surface. They can only occur near the poles. Such eclipses can still be total or annular, and occur near the beginning or end of a Saros series. Although many occur in our 3,000-year period, only three are visible in their annular form from the UK – the eclipses of 2003, 2662 and 2676.

CONTACTS The terms first contact, second contact, third contact and fourth contact are used to describe the various stages in a solar eclipse. The terms are defined on page 21. The terms have a slightly different meaning depending on whether they are referring to an annular or a total eclipse. Except on rare occasions near the poles, the Moon crosses the Sun from right to left.

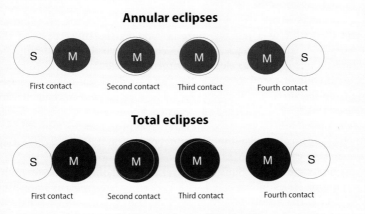

Annular eclipses

First contact   Second contact   Third contact   Fourth contact

Total eclipses

First contact   Second contact   Third contact   Fourth contact

In the diagram the Moon moves from right to left

eclipse will be displaced from its 18-year-old sister by about 11 days and one-third of the world to the west. Successive solar eclipses in a Saros series also gradually shift their tracks northwards or southwards. Odd-numbered Saros series shift southwards, and even-numbered shift northwards on the Earth.

Exeligmos After three Saroses (54 years 33 days; a period known as an exeligmos), eclipses return to roughly the same position on Earth, shifted slightly northwards or southwards. For a detailed explanation of the Saros, see: w eclipse.gsfc.nasa.gov/SEsaros/SEperiodicity.html#section104.

ANATOMY OF THE SUN Our knowledge of the Sun has grown significantly in recent years, partly because of the spacecraft SOHO, which creates a permanent eclipse between itself and the Sun, allowing constant monitoring of its outer layers.

The core of the Sun is a furnace of burning hydrogen at 15 million °C, hidden from view by a shell of opaque gases 696,000km thick. Working outwards, through layers of radiation and convection, we reach the **photosphere**. This is the Sun's visible surface – a boiling sea of rising and sinking gases at 5,500°C. It is the layer where sunspots emerge and vanish. The photosphere is only about 300km deep.

Above this layer is the **chromosphere**, a vivid orange colour, just 2,500km thick. Its temperature varies strangely, the lower levels being about 4,000°C and the upper layers hotter, at about 10,000°C. The chromosphere is not smooth but covered with sporadic projections of gas which can reach 700km high, as far as the next layer – the corona.

The **corona** is a haze stretching far from the Sun, growing ever thinner yet still detectable as a tenuous haze beyond Earth. Those parts of the chromosphere which stretch into the corona soar in temperature to 1 million °C. This heat accelerates charged particles outwards, imbuing them with enough speed to escape the Sun's gravitational field. The gases rising from the surface of the Sun expand as they go, and so their density decreases and they are more easily warped by the Sun's magnetic fields. This twists them into loops and arches known as **prominences**. Sometimes a magnetic loop breaks and billions of tonnes of gas shoot into space, creating a coronal mass ejection (CME).

X-ray images reveal the Sun's outer layer to be full of roller-coaster loops and exploding gases interspersed with darker, quieter regions known as **coronal holes**. The beauty of a total solar eclipse is that, with the glare of the photosphere obscured, you can for once witness the chromosphere, prominences and corona.

EFFECTS SEEN DURING A TOTAL SOLAR ECLIPSE Although the arrangement of the Sun, Moon and Earth may seem identical, many factors combine to change the look of every eclipse. For example:

- How far the Sun is above the horizon
- The amount of cloud and dust in the sky
- The difference between the sizes of the Sun and Moon
- The libration (oscillating motion; see below) of the Moon
- Where the Sun is in its activity cycle

These factors manifest themselves in the shape and size of the corona, the landscape, shadow bands, and the location of Baily's beads and the diamond ring.

Libration Libration is the apparent wobble or variation in the visible side of the Moon that permanently faces the Earth, allowing observers on Earth to see, over a period of time, slightly more than half of the lunar surface.

As the wobble changes, different mountain ranges and valleys appear at the Moon's limb. Valleys allow the Sun to shine through and produce Baily's

Ns = solar north axis; Nm = lunar north axis; C2 = 2nd contact; C3 = 3rd contact; Zenith = observer's zenith

beads and the diamond ring (see below). Because the wobble is precisely known, we can predict where the valleys and mountains can be by producing a Lunar Limb Profile (LLP). LLPs have to be produced for the exact date, time and location and these are given for each totality in this guide.

On page 17 is a sample LLP. The most important things to look for are the valleys at C2 (beginning of totality) and C3 (end of totality).

The chart implies that you should look for a diamond ring at the start of totality at roughly the 11 o'clock position, and a smaller diamond ring should appear at roughly halfway between the 4 and 5 o'clock positions at the end of totality. The chart is arranged so that when you are looking at the eclipse the zenith (the overhead point) is pointing straight up. The scales in the middle are of interest only to those who need to know the precise orientation of the Sun and Moon.

The corona During periods of solar minimum, the corona is roughly arranged in the Sun's equatorial regions. However, during the Sun's active periods, the corona is more evenly distributed, though it is most prominent in areas with sunspot activity. The solar cycle spans approximately 11 years. See page 20 for photographs showing solar minimum (1) and solar maximum (2). (Solar minimum years are 1953, 1963, 1975, 1987, 2009, 2020; solar maximum years are 1958, 1969, 1980, 1990, 2001, 2014, 2024 predicted.)

Baily's beads The Moon does not have a smooth outline, and as the Sun is just about to be covered (second contact) its last vestiges shine through the lunar valleys. Francis Baily was the first to discuss these and they are now known as Baily's beads. To predict in advance where beads and the diamond ring will occur on the Sun's disc, you need to refer to the relevant LLP chart. The location of Baily's beads is different for every eclipse and their location changes depending on where in the eclipse track you observe. The LLPs have been provided for the start, middle and end locations for each total eclipse. C2 and C3 mark the points of second and third contact. If you were located near the edge of the eclipse track, you would witness extended duration beads at the north and south positions. If the Moon and Sun are almost exactly the same size you can get beads all around.

Diamond ring When all the beads have disappeared except the last one, this is known as the diamond ring. It should also appear exactly as totality

◀ *Taken on Hikueru Island, French Polynesia in July 2010, showing diamond rings at* **1** *the start &* **2** *the end of totality, through thin cloud.*

2008 **1**

2001 **2**

3

4

5

ends. Photographs 3 and 4, opposite, show the beginning and end diamond ring and were taken in Libya in 2006.

On examining the LLP, if you see a deep valley near C2 and C3 this should result in a large diamond ring at the start and/or end of totality.

Shadow bands Shadow bands are thin wavy lines of alternating light and dark that can be seen moving and undulating in parallel on plain-coloured surfaces immediately before and after a total solar eclipse. Shadow bands result from the illumination of the atmosphere by the thin solar crescent a minute or so before and after totality. These can be quite elusive and difficult to see.

In 1842, George B Airy, the Astronomer Royal, saw his first total eclipse of the Sun and recalled shadow bands as one of the highlights:

> As the totality approached, a strange fluctuation of light was seen upon the walls and the ground, so striking that in some places children ran after it and tried to catch it with their hands.

WHAT TO WATCH FOR AS THE ECLIPSE UNFOLDS: STEP BY STEP

> An unearthly gloom enveloped us, and the atmosphere grew chill, so that we shivered perhaps as much with awe as with the cold…
> R A Marriott, *A British Eclipse*, 1927

Once you are well inside the path of totality (which you can work out using the maps and tables in this booklet) you can settle down to watch the spectacle unfold. Do not worry too much about memorising the precise timings of the stages of the eclipse – as long as you know roughly (to the nearest 15 minutes) when to look, nature will guide you through the experience.

When the Moon first encroaches on the Sun, the moment known as **first contact** has arrived. It is followed roughly 1 hour later by **second contact** – the moment when the Moon drifts fully in front of the Sun and totality begins. **Third contact** heralds the end of totality because the Moon starts to shift away from the Sun. **Fourth contact** is when the two discs finally part.

At first an eclipse is unnoticeable; it's just a shallow scoop from one edge of the Sun which hardly diminishes the daylight. The fading of the day happens very slowly and, until the Sun is over 80% covered, there is little more than a faint dullness. Even when the Sun is 80% covered, most

◀ *The corona during period of* **1** *solar minimum &* **2** *solar maximum. The diamond ring at* **3** *the start &* **4** *end of totality (Libya, 2006).* **5** *Solar prominence with Baily's beads (China, August 2008).*

First contact: *The moment when the edge of the Moon first encroaches on the Sun.*

The Moon approaches 50% coverage of the Sun but the ambient light has changed little.

Second contact: *The diamond ring: a brilliant gleam of light appears as the last bit of the Sun shines through a lunar valley.*

Totality: *Mid eclipse: the Moon completely covers the Sun.*

Third contact: *The Sun emerges through a lunar valley, totality has ended.*

Fourth contact: *The Moon takes its leave of the Sun. The eclipse is over.*

The direction of the Moon's travel varies depending on your location.

of us will notice little change in the light because our brains are so used to compensating for the effects of heavy cloud. But the Moon crawls relentlessly onwards and, when it has covered 90% of the Sun, the light will fail noticeably and the land will turn a bluish grey. With about 10 minutes to go, things start to happen at an increasing rate.

From about 5 minutes before totality you should start watching for brighter planets such as Mercury and Venus near the Sun. During the last few minutes before second contact, daylight will disappear fast. Now is the moment to watch the ground for **shadow bands** rippling across lightly coloured surfaces, as if on a lake. The ripples appear because of a lensing effect caused by the Earth's atmosphere which bends and focuses the light streaming towards Earth from the Sun's thin remaining crescent. You can see them more easily if you lay a white sheet on the ground. Note also tiny crescent Suns projected onto the ground through gaps between the leaves of trees and bushes. You may like to take a piece of card punctured with holes (using a knitting needle or similar) in case there is no vegetation to do the job for you.

At around this time, and assuming that it is not already a windy day, you may be aware that there is a distinct breeze. This can be sufficient to blow any paperwork around that may be lying about. This was thought to be a coincidence, but can be easily explained because the air inside the Moon's shadow is cooler (and hence of different density) than the air outside the shadow. This causes air movement and hence wind. The wind was very noticeable in the Libyan desert in 2006, and also in the Australian outback in 2002.

About 15 seconds before the Sun is completely eclipsed, you will see specks of light appear around the dark disc of the Moon, just like a string of pearls. These are **Baily's beads** (page 19; see also photo 5 on page 20), the last few rays of sunlight shining through valleys on the edge of the Moon. They will swiftly die away until just one remains – a dazzling jewel known as the **diamond ring** (page 19), which shines brilliantly for just three or four seconds and then vanishes.

From the moment of the diamond ring's appearance it is safe to watch the eclipse with the naked eye, though you may pay a penalty for viewing the diamond by being unable at first to distinguish the features of the ensuing totality. You will know when totality has arrived. Suddenly the Moon's shadow rushes across the land towards you and your world is plunged into twilight. At this point the emotion can be overwhelming.

Now the secrets of the Sun's outer layers, usually hidden by the glare of its fiery centre, are briefly revealed. For the first few seconds of totality you may see a vibrant, pinkish-red rim at the edge of the Sun. This is the light from the Sun's lower atmosphere, the **chromosphere**. As the Moon moves onwards, it will swiftly cover this too (during very short eclipses the chromosphere never really disappears because the Moon is only minutely larger than the Sun).

An effect that should last a little longer will be the appearance of several deep red clouds, like smoke drifting from the surface of the Sun. These are the solar **prominences**, stretching to a distance of up to one-twentieth of the Sun's diameter.

At full totality you will see the black disc of the Moon perfectly surrounded by the pearly light of the Sun's corona. Wispy plumes and streams of coronal light will dance outwards.

If you can drag your eyes away from the eclipse itself, you will have an unprecedented opportunity to view the daytime positions of planets and stars which began to emerge just before totality.

Don't forget to check the landscape, where nature will respond as if night has fallen. Flowers will close their petals and disorientated bees will cease their flight. Insects, maintaining their daytime silence until the moment of totality, will suddenly sing as if it were night, and mosquitoes will start biting. Animals that were grazing peacefully may become nervous and confused, while the birds above them will flock into the trees with great fuss. Cows have been known to line up and walk to their shelters.

In the distance, where the eclipse is only partial, the horizon will be an evening pink. The temperature plunge will be noticeable.

If you are near the centre line, the diamond ring emerges again at roughly the opposite edge of the Sun from before. Totality has finished and the sequence will begin again in reverse. At the diamond ring's appearance look away or use your eclipse viewer. At fourth contact, the eclipse is over.

If the day is thick with cloud, then most of these sights will be hidden, but partial cloud and thin cloud will not necessarily spoil the experience. The eclipse carries with it its own weather patterns, caused by the wild temperature variations that the moving shadow creates on Earth. This can cause clouds miraculously to disperse during the seconds leading up to totality – but more often it can cause them swiftly to materialise out of a previously clear sky (rapid cooling of the air can have either effect, depending on the humidity and pressure). Trying to catch a glimpse of totality through the spaces between thick clouds has its own thrills – even one fleeting view seems a tremendous achievement and is deeply exciting. If there is broken cloud, you may see patterning and colours in the sky. If there is thin cloud, the corona may still be visible. Finally, for those besieged by the weather, watch out for a shadow that the Moon may cast onto the cloud above you.

CHOOSING WHERE TO SEE AN ECLIPSE Many people think that the only place to go to see an eclipse is on the centre line as this maximises the duration of totality. However, there are many things to consider when choosing your location.

Weather Simple: choose the least cloudy location. Be aware also of the possibility of pollution, and smoke from forest fires and local chimneys. Sandstorms can occur in some areas too. Do not choose a location near to roads where car headlights can spoil the picture. Beware of hotel floodlights that may come on as darkness descends.

1 *Setting up on the beach to view the eclipse, Anaa Island, French Polynesia (July 2010).* 2 *Solar eclipse captured in the path of totality in the mountains of Stanley, Idaho, USA (21 August 2017).* ▶

Wildlife Animals react to large eclipses and you may want to witness their behaviour.

Scenery Deserts have a stark beauty, and are a natural choice for low cloud cover. But foliage gives special effects by projecting the Sun through its leaves. How about photographing the eclipse through a rock arch, or with a tree in the foreground?

Sun's altitude Many people head for the location of maximum eclipse where the Sun is highest in the sky as this can reduce the chances of cloud and increase the duration of totality. Although this makes sense, you should not rule out observing where the Sun is lower, either just after sunrise or before sunset. Some of the most dramatic eclipses occur at sunrise and sunset because the Sun/Moon always appear larger when framed against objects in the foreground. This provides the opportunity to take amazing photographs of the eclipse though rock arches, bridges or buildings.

Below are sunset locations and the Sun's altitude for eclipses described in this book:

- **2023 annular** 17° Pacific coast of Oregon, USA
- **2026 total** 2° Balearic Islands
- **2027 annular** <5° Ghana, Togo, Benin
- **2028 annular** 2° Valencia and Balearic Islands

Eclipse duration Maximum duration is not necessarily on the centre line. The Moon's mountainous rim can make maximum totality occur slightly away from the centre line. Also consider that in eclipses with a duration of around 2 minutes or more, you can be 20km from the centre line and still only lose a few seconds of totality.

Eclipse effects There are people who choose to observe totality near the northern or southern limits. They sacrifice the length of totality to see extended Baily's beads and prominences.

CLIMATE/WEATHER Climate/weather details are included for each of the total eclipses in this guide.

Travelling the world to see solar eclipses always comes with the risk of disappointment owing to the vagaries of weather. Even for locations where climate statistics look very promising, showing that the sun will shine on the big day, the bottom line is that it's the weather that you get and not the statistics of climate!

The best approach is to watch weather forecasts in the days leading up to the eclipse and be prepared to relocate if necessary. Additionally, satellite imagery, which is widely available and free on the internet, is an invaluable source from which to make informed decisions if relocation is required. In some instances, it may simply be moving a few kilometres to avoid patches of localised cloud; but in the worst case it could mean travelling hundreds of kilometres to avoid large-scale weather systems.

ECLIPSES IN HISTORY

> The Sun...
> In dim eclipse disastrous twilight sheds
> On half the nations, and with fear of change
> Perplexes monarchs.
>
> John Milton, *Paradise Lost*, Book I, lines 587–600 (1667)

Milton's sentiment remains true today. A total solar eclipse bypasses the brain and its scientific understanding, and speaks primitively to the heart. It reminds us of our cosmic insignificance.

Eclipses caused far more perturbation when they were unpredictable – and historians have studied ancient responses to them.

One of the oldest known eclipse tales comes from China and relates to two astronomers, Hsi and Hoe, who became drunk and failed to respond adequately to a total solar eclipse, a crime which was punished by death. If this were true, it would mean that the first recorded eclipse occurred between 2159 and 1948BC. But it is almost certainly a myth, perhaps a morality tale aimed at civil servants of the time.

The first person to predict totality with success was thought for a long time to have been Thales of Miletus. He is said to have foretold the year and position of an influential eclipse that plunged a battlefield into darkness during an encounter between the Lydians and the Medes. Chastened by the experience, the two sides laid down their arms and agreed a peace deal. Today's astronomers cannot see how Thales could have predicted an eclipse given the limited knowledge at the time. It is generally agreed that, while the solar drama may well have ended a war, probably on 28 May 585BC, it was not actually predicted.

Ancient civilisations such as the Egyptians, the Chinese and the Babylonians all developed methods for tracking the motions of the Sun and Moon, and had accurate constellation maps. The best-known eclipse archive was developed by the Chaldeans who began to keep precise historical records from 750BC to AD75. They developed mathematical theories from these records and, in the 4th century BCE (the Hellenistic period), they made the breakthrough of examining past records to look for repeating cycles: eclipse prediction had begun.

Even today, those Babylonian records, stored on thousands of clay tablets, are useful. In 1997, F Richard Stephenson, an astronomer from Durham University in the UK, used a discrepancy between early eclipse records and today's calculations to show that the Earth's rotation is slowing by 2.3 milliseconds per century. His work was published in *Historical Eclipses and Earth's Rotation* (page 99).

Scientific study of the Sun during eclipses began in the early 19th century with Francis Baily, who gave his name to the 'string of bright beads' that fringes the Moon just before totality. In 1715, Edmund Halley (of comet fame) produced a paper for the general public encouraging them to look at the May eclipse that crossed London and large parts of Britain. This was Britain's last really good total eclipse and it lasted more than 4 minutes in a cloudless May sky. Eclipses were studied intensely because they allowed astronomers to determine the distance of the Sun and Moon with greater accuracy. An early 20th-century eclipse even helped prove Einstein's theory of relativity.

After these successful public relations activities, scientists began their modern pursuit of eclipse chasing, pointing all manner of instruments at the Sun to exploit the rare moment when its faint corona could be viewed unimpeded.

2

The Total Solar Eclipse of 8 April 2024

WHERE IT GOES *Figure 2.1*

The path of the Moon's umbral shadow starts in the Pacific Ocean, crosses Mexico, the USA and Canada, ending in the north Atlantic. For a detailed map of the track, see **w** xjubier.free.fr/tse2024map.

Figure 2.1

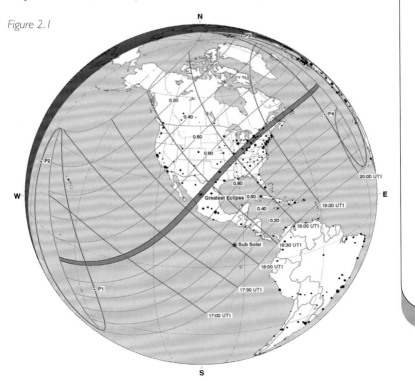

For this eclipse the weather tends to be significantly better across Mexico than the USA or Canada, as shown in the average cloud chart below (Figure 2.2).

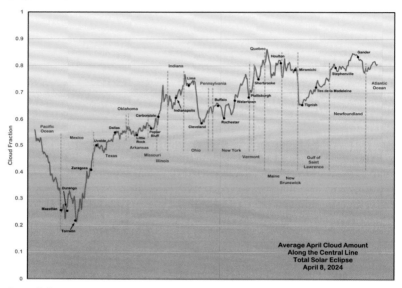

Figure 2.2

MEXICO By the time the lunar shadow reaches the coast of Mexico, it is already nearly 7,000km and 90 minutes old. The shadow's introduction to Mexico is at Isla Socorro, a small volcanic outcrop lying 680km west of the mainland, and the first of a half-dozen small islands before the continent. The marine environment surrounding the islands guarantees a relatively high cloud cover, but which declines rapidly as the eclipse track approaches the mainland beaches near Mazatlan.

On the coastal strip around Mazatlan, convective clouds are uncommon and southward-sweeping cold fronts make little impact: the frontal cloud and moisture are blocked by the heights of the Sierra Madre Occidental. Most cloud on the Pacific side of the mountains comes from jet-stream cirrus and mid-level cloud, similar to further inland. While the Mazatlan coast doesn't have the greatest sunshine along the eclipse track, it's only 3–5% cloudier than Durango and Torreón.

Morning fog or low stratus clouds that form offshore overnight and come ashore along the coast may also be a problem, as they tend to be drawn inland as the Sun rises. The fog or stratus will burn off or thin later in the day, but that may not happen before eclipse time and there is a risk that the fog will re-form as the air cools ahead of totality. The only reliable way to escape the fog is to move away from the coast and uphill to altitudes of 100–200m, but stratus clouds can reach much further – a distance of 50km from the coast.

The Mexican coast presents an abrupt mountain barrier to the sea and so the eclipse path has only a narrow 25km coastal plain to cross before beginning to rise over the Sierra Madre Occidental; within 100km of the coast, the terrain has risen to over 2,600m. Once across the Sierra, near Durango, the Moon's path enters onto the Mexican Plateau, a rough inland mesa consisting of desert plains interspersed with low mountain ridges. North of Torreón, the shadow track passes over the Bolson de Mapimf, a large ill-defined inland basin that stretches north to Presidio on the US border.

The dry-season climatology of Northern Mexico in April promises a generous probability of sunny weather for the eclipse, though the presence of thin cirrus-level clouds may occur. In satellite measurements, centre line cloudiness barely budges from the 25–35% range from Mazatlan to the start of the Sierra Madre Oriental. Only when the track reaches the east side of the Orientals and descends onto the Gulf Plains does the cloud cover rise, from around 30% at Monclova to near 50% at Piedras Negras and the US

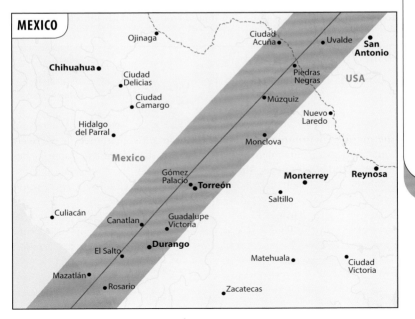

border. Though Mexico has a very sunny climate, note that approximately one-quarter to one-third of skies are cloudy enough to have an impact on viewing the eclipse.

USA AND CANADA As we look along the graph of centre line cloudiness, we see a modest increase in cloud cover from the Mexican border to the Missouri River. Most of the rise takes place in Texas.

The best Texas weather prospects, and the best for the USA and Canada, lie on the Edwards Plateau, where median cloud amounts are as much as 15% lower than on the Coastal Plain.

Cloud amounts climb slowly along the shadow path, rising from 47% to 56%. Through Oklahoma, Arkansas, and most of Missouri, cloud cover along the track axis varies between 54% and 59%. It is not until the eclipse shadow is almost in Illinois that cloud amount reaches above 60%.

Until the eclipse track passes Ohio, typical April weather can include a threat of severe thunderstorms, including tornadoes, though they usually hold off until later in the afternoon, after the shadow passage.

On the centre line, cloud amounts vary between 60% and 70% from Illinois to Lake Erie. Just beyond Sandusky, Ohio, centre line cloud drops abruptly by more than 15% as the track reaches and crosses Lake Erie. Low-cloud refuges are created by the lakes, and they are found above the lakes and for a short distance along and inland from the south shores.

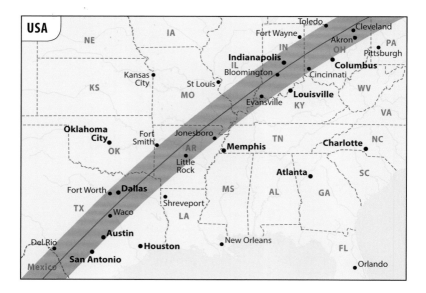

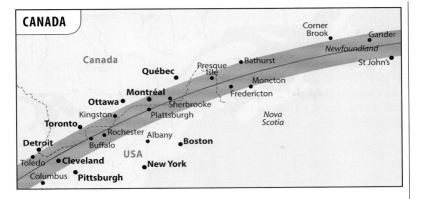

As the umbral shadow moves over Eastern Ontario and the Northeast States, it begins to cross the northern reaches of the Appalachian Mountains and terrain-induced cloud changes become significant. Lake Ontario's influence is reflected in an 8% decline in cloudiness where the centre line crosses the shore at Rochester, which shows that Lake Ontario is as effective as Lake Erie and Lake St Clair in moderating cloudiness. The impact of the Adirondack Mountains is particularly noticeable in the centre and south-limit curves, with an increase of about 10% in the cloud cover.

As the shadow path crosses into Vermont, cloud amounts climb. Along the Maine–Quebec border, cloudiness reaches 90% across the White and Longfellow mountains and barely falls below 75% over the rest of the path through eastern Maine and over New Brunswick.

Cloudiness shows an abrupt drop as the eclipse reaches the Northumberland Strait on the northeast coast of New Brunswick. The largest part

2024 TOTAL ECLIPSE FACTS

Time of greatest eclipse 18:17:18 GMT
Maximum duration of totality 4m 28s
Location of greatest eclipse 25°17.4′N, 104°08.3′W
Altitude of Sun at greatest eclipse 70°
Maximum width of path of totality 197.5km
Saros Series 139, member 30 of 71
Magnitude 1.0567
Obscuration 1.1163
Gamma 0.3431

of this drop is over the Strait itself, extending only up to the land's edge or a very short distance inland. The advantage gained by settling along the Northumberland shore is only about 15%.

After crossing the Northumberland Strait, the Moon's shadow passes Prince Edward Island, and heads across the Gulf of St Lawrence. Cloud amounts fall abruptly to just under 65% when the eclipse track moves off the New Brunswick coast and out onto the waters of the Gulf. The whole of the shoreline from North Cape to West Point benefits from the cold water

LOCAL ECLIPSE CIRCUMSTANCES, 2024

The following pages contain information on timings for locations from west to east. If your location is not mentioned, you should interpolate between nearby locations. All times are GMT.

Location	1st	2nd	Mid	3rd	4th	Dur/%	Alt
CANADA							
Charlottetown, PE	18:28:06	–	19:38:14	–	20:43:55	0.992	32
Fredericton, NB	18:23:39	19:33:46	19:34:55	19:36:04	20:41:52	2m 17s	35
Halifax, NS	18:27:16	–	19:38:00	–	20:44:06	0.946	33
Hamilton, ON	18:03:53	19:18:10	19:19:06	19:20:02	20:31:09	1m 51s	46
London, ON	18:01:39	–	19:17:09	–	20:29:41	0.995	47
Montreal, QC	18:14:26	19:26:53	19:27:31	19:28:09	20:36:50	1m 15s	40
Ottawa, ON	18:11:33	–	19:25:02	–	20:35:00	0.985	42
Quebec, QC	18:18:16	–	19:30:13	–	20:38:20	0.983	38
Saint John's, NF	18:39:29	–	19:45:55	–	20:47:53	0.989	23
S St Marie, ON	18:01:55	–	19:15:04	–	20:25:58	0.861	46
Toronto, ON	18:04:56	–	19:19:54	–	20:31:38	0.997	45
Windsor, ON	17:58:21	–	19:14:21	–	20:27:38	0.992	49
USA							
Albany, NY	18:12:13	–	19:26:35	–	20:36:54	0.966	42
Austin, TX	17:17:11	18:36:10	18:37:00	18:37:50	19:58:06	1m 39s	67
Boston, MA	18:16:05	–	19:29:47	–	20:39:07	0.931	40
Buffalo, NY	18:04:55	19:18:19	19:20:12	19:22:04	20:32:08	3m 45s	46
Chicago, IL	17:51:22	–	19:07:36	–	20:21:58	0.942	52
Cincinnati, OH	17:52:19	–	19:09:55	–	20:24:59	0.994	52
Cleveland, OH	17:59:19	19:13:44	19:15:38	19:17:32	20:28:58	3m 48s	49
Columbus, OH	17:55:35	–	19:12:43	–	20:27:02	0.996	51
Dallas, TX	17:23:16	18:40:40	18:42:37	18:44:33	20:02:39	3m 53s	65
Detroit, MI	17:58:20	–	19:14:19	–	20:27:36	0.991	49
Hartford, CT	18:13:17	–	19:27:37	–	20:37:43	0.929	42
Houston, TX	17:19:59	–	18:40:11	–	20:01:08	0.943	67
Indianapolis, IN	17:50:32	19:06:02	19:07:57	19:09:52	20:23:11	3m 49s	53

of Northumberland Strait, with cloud amounts of just over 65%. These are slightly better than the shoreline on the opposite side of the Strait, in New Brunswick.

Beyond Prince Edward Island, the cloud cover climbs inexorably upward, from the minimum of 65% near Tignish to a discouraging 85% near Gander.

Clear-sky prospects are limited in Newfoundland with winter only beginning to give way to the spring warming.

Location	1st	2nd	Mid	3rd	4th	Dur/%	Alt
Louisville, KY	17:49:05	–	19:07:09	–	20:22:56	0.989	54
Memphis, TN	17:37:41	–	18:56:58	–	20:15:03	0.976	60
Nashville, TN	17:44:33	–	19:03:23	–	20:20:13	0.949	57
New York, NY	18:10:33	–	19:25:32	–	20:36:22	0.911	43
Oklahoma City	17:27:23	–	18:45:29	–	20:04:09	0.94	62
Pittsburgh, PA	18:00:48	–	19:17:19	–	20:30:30	0.971	48
Providence, RI	18:15:17	–	19:29:13	–	20:38:47	0.918	41
Rochester, NY	18:06:59	19:20:07	19:21:57	19:23:46	20:33:24	3m 38s	44
San Antonio, TX	17:14:28	–	18:34:18	–	19:55:43	0.998	68
St Louis, MO	17:42:56	–	19:00:48	–	20:17:21	0.988	56
MEXICO							
Chihuahua	17:03:02	–	18:20:28	–	19:41:16	0.918	66
Durango	16:55:13	18:12:07	18:14:00	18:15:53	19:36:39	3m 51s	70
Guadalajara	16:50:34	–	18:09:31	–	19:32:36	0.913	73
Guadalupe	17:04:42	–	18:24:34	–	19:47:02	0.952	71
Matzalan	16:51:23	18:07:25	18:09:34	18:11:43	19:32:00	4m 15s	69
Matzalan centre line	16:51:15	18:07:16	18:09:26	18:11:43	19:32:06	4m 27s	70
Monterrey	17:04:34	–	18:24:25	–	19:46:54	0.954	71
Nazas	16:58:24	18:15:05	18:17:18	18:19:31	19:39:45	4m 25s	70
Torreon	16:59:53	18:16:52	18:18:58	18:21:04	19:41:24	4m 11s	70
Zapopan	16:50:34	–	18:09:30	–	19:32:34	0.916	73
Zaragoza	17:09:11	18:26:23	18:28:36	18:30:50	19:50:20	4m 27s	69

1st The instant when the Moon first touches the Sun

2nd The true start of the totality when the diamond ring and Baily's beads cease. If no time is given, then the eclipse is not total at that location.

3rd The true end of totality at the instant that Baily's beads reappear. If no time is given, then the eclipse is not total at that location.

4th The instant when the Moon is no longer visible on the Sun

Dur Duration of totality in min:sec. If a percentage is given, this is because the eclipse is not total at that location.

Alt The Sun's height above the horizon

SKY CHART FOR THE MEXICO COAST, 2024 *Figure 2.3*

Mercury and Venus should be easily visible before totality, together with fainter Jupiter, Saturn and Mars during totality. Stars that may be visible are also shown.

LUNAR LIMB PROFILE FOR THE MEXICO COAST, 2024 *Figure 2.4*

Look for a diamond ring at the start of totality at roughly the 10 o'clock position, and a diamond ring at roughly halfway between the 2 o'clock and 3 o'clock positions at the end of totality.

SKY CHART FOR CENTRAL USA, 2024 *Figure 2.5*

Mercury and Venus should be easily visible before totality, together with fainter Jupiter, Saturn and Mars during totality. Stars that may be visible are also shown.

Figure 2.3

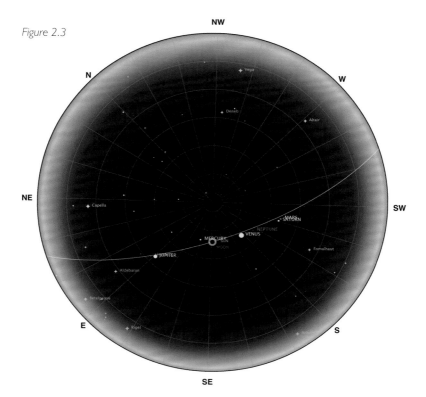

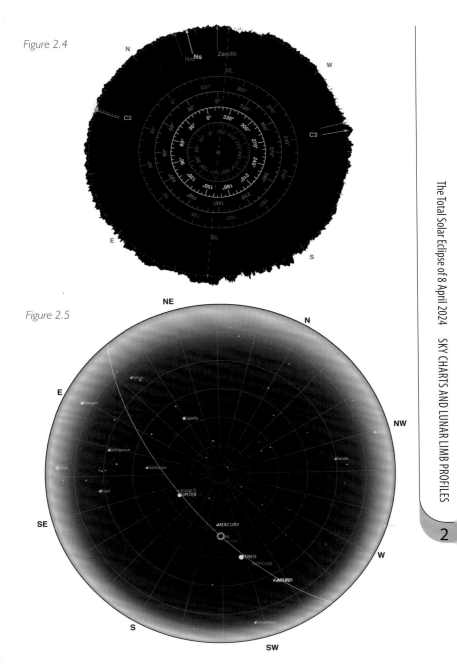

Figure 2.4

Figure 2.5

LUNAR LIMB PROFILE FOR CENTRAL USA, 2024 *Figure 2.6*

Look for an extended diamond ring at the start of totality at roughly the 11 o'clock position, and a diamond ring at roughly the 5 o'clock position at the end of totality.

SKY CHART FOR EAST USA AND CANADA, 2024 *Figure 2.7*

Mercury should be easily visible before totality together with Venus very low on the western horizon. Fainter Jupiter should be visible during totality. Stars that may be visible are also shown.

LUNAR LIMB PROFILE FOR EAST USA AND CANADA, 2024 *Figure 2.8*

Look for a diamond ring at the start of totality at roughly halfway between the 11 o'clock and 12 o'clock positions, and a very much larger diamond at roughly the 5 o'clock position at the end of totality.

Figure 2.6

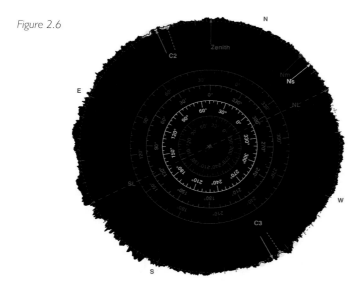

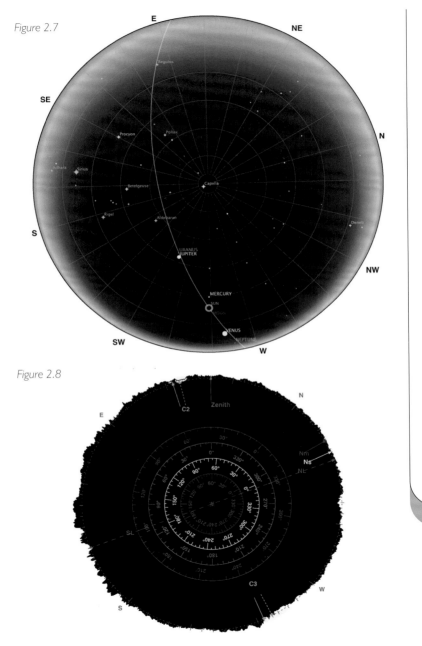

Figure 2.7

Figure 2.8

THE SAROS

The total eclipse of 2024 is the thirtieth member of 71 eclipses in Saros series 139. All eclipses with an odd number occur at the Moon's ascending node and the Moon moves southwards with each member in the family. The series started on 17 May 1501 and includes 71 eclipses. It begins with 16 partials, followed by 0 annulars, then 55 totals, and finally ends with nine more partials. The series ends on 3 July 2763.

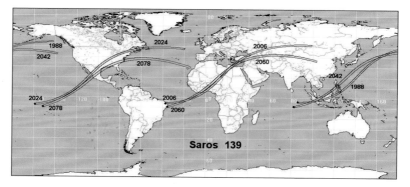

Saros 139 contains the longest total eclipse on record of 7m 29s on 16 July 2186. It contains five eclipses with durations longer than 7 minutes!

The map above shows the 2024 eclipse track together with the previous and next eclipses in the Saros. Notice they shift around the globe, moving west by 8 hours each time.

3

The Total Solar Eclipse of 12 August 2026

The total solar eclipse of 2026 starts in the northern tip of Russia near the Laptev Sea before crossing near the North Pole and into Greenland. From there it enters eastern Iceland including Reykjavik. It enters Spain's north

Figure 3.1

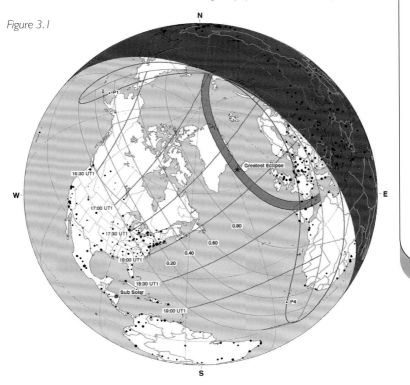

coast crossing northern Spain (the northern part of Madrid will witness totality) and then the Balearic Islands. For a detailed map of the track, see w xjubier.free.fr/tse2026map.

CLIMATOLOGY

ICELAND Iceland is affected by Atlantic depressions all year round and consequently has a wet and cloudy climate. According to statistics for Reykjavik, where totality passes, the average amount of August sunshine is 155 hours, which corresponds to the sun shining just 39% of the total time possible.

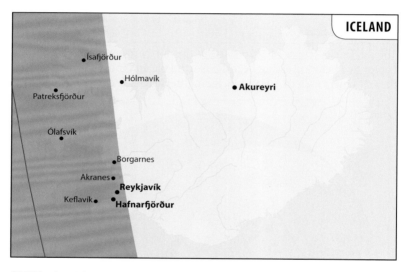

SPAIN Thankfully, as the total solar eclipse crosses Spain in August 2026, it does so during one of the country's sunniest months. However, there are regional differences in the climate where the eclipse umbral shadow crosses which should be considered to maximise your chance of a successful observation.

Many seasoned 'eclipse chasers' (page 92) are likely to consider the northwestern part of the country, where the Moon's umbral shadow first makes landfall with Spain. The benefit here is that the eclipse is higher in the sky. If located on the coast, you should get good views across the sea when the Moon's umbral shadow races in before totality. Unfortunately, statistics show that this location offers the greatest chance of the event being interrupted by cloud. Although this region of Spain offers good amounts of sunshine in summer, the maritime influence means there is

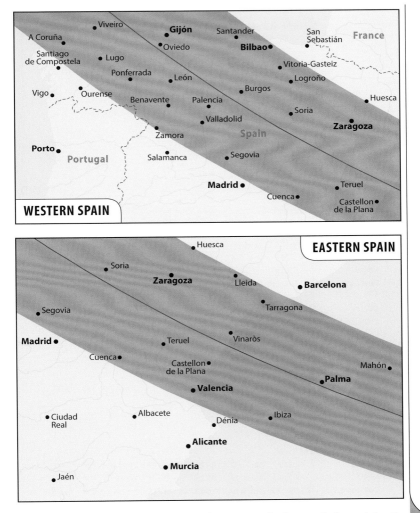

always a distinct possibility for cloud to encroach the north from Atlantic low-pressure areas.

However, this scenario will be quashed if the Azores' high-pressure area successfully builds in from the southwest, thus minimising the threat; unfortunately this is an intermittent feature and can never be relied upon. A large part of the northern coast is bordered by the Cantabrian Mountains, which exacerbate cloud and rainfall in the region, further limiting potential sunshine. Long-term statistics show that rainfall

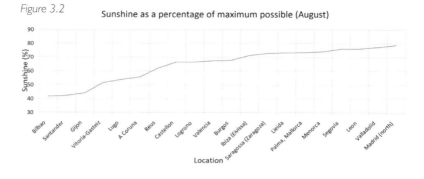

Figure 3.2

Sunshine as a percentage of maximum possible (August)

LOCAL ECLIPSE CIRCUMSTANCES, 2026

Location	1st	2nd	Mid	3rd	4th	Alt	Dur/%
ICELAND							
Patreksfjordur	16:43:41	17:44:27	17:45:29	17:46:31	18:44:44	25°	2m 04s
Reykjavík	16:47:08	17:48:11	17:48:42	17:49:13	18:47:34	24°	1m 03s
UK							
Edinburgh	17:08:35	–	18:05:46	–	19:00:02	14°	92%
London	17:17:16	–	18:13:17	–	19:06:17	10°	93%
Penzance	17:18:58	–	18:16:18	–	19:10:19	13°	96%
IRELAND							
Cork	17:15:01	–	18:13:18	–	19:08:14	15°	96%
FRANCE							
Paris	17:22:10	–	18:17:16	–	19:09:23	8°	93%
SPAIN & PORTUGAL							
Oviedo	17:31:17	18:27:02	18:27:56	18:28:51	19:21:02	10°	1m 48s
Bilbao	17:31:43	18:27:17	18:27:33	18:27:48	19:19*	8°	31s
Valladolid	17:34:26	18:29:49	18:30:32	18:31:15	19:23*	9°	1m 27s
Zaragosa	17:34:37	18:28:58	18:29:40	18:30:23	19:08*	6°	1m 23s
Barcelona	17:35:00	–	18:29:13	–	18:55*	4°	99.7%
Madrid, Alcobendas	17:36:28	18:31:57	18:32:06	18:32:15	19:17*	7°	15s
Madrid central	17:36:43	–	18:32:20	–	19:17*	7°	99.8%
Palma	17:37:59	18:31:00	18:31:48	18:32:36	18:50*	2°	1m 36s
Valencia	17:38:19	18:32:25	18:32:54	18:33:24	19:02*	4°	1m 0s
Ibiza	17:39:10	18:32:40	18:33:12	18:33:44	18:53*	2°	1m 4s
Lisbon	17:39:13	–	18:36:03	–	19:35*	10°	95%

*Sun sets before end of eclipse

Time of greatest eclipse 17:45:53 UT
Maximum duration of totality 2m 18s
Location of greatest eclipse 65°13.0′N, 25°13.6′W
Altitude of Sun at greatest eclipse 26°
Maximum width of path of totality 293km
Saros Series 126, member 48 of 72; only one more total eclipse follows in 2044, the remainder are partials.
Magnitude 1.0386
Obscuration 1.0788
Gamma 0.8977

amounts vary from 46mm, falling over nine days at A Coruña on the Atlantic coast, to 84mm falling over 14 days in the more mountainous areas such as Santander.

The interior of Spain, away from the tempering Atlantic influence, becomes very hot in summer and, consequently, daily sunshine totals are greatly increased (Figure 3.2). Statistics show that during August, Madrid, whose northern suburbs are on the limits of totality, receives 335 sunshine hours with 14 predominately sunny days.

Intense daytime heat always has the potential to generate thunderstorms. If a thundery outbreak does occur (as in 2018) near the time of the eclipse, then the cloud shield generated would obscure the sky over a large area. Satellite imagery also demonstrates a risk, albeit a low one, of decaying weather fronts (as in 2015) pushing into the interior of the country from the Atlantic.

Since the eclipse occurs when the Sun is at a relatively low altitude in the sky, even if the sky were filled with broken cloud, the gaps between the clouds will appear to decrease in size when looking nearer to the horizon, thus reducing the chance of a successful observation.

The path of totality leaves the mainland and crosses the well-known Mediterranean island resorts of Ibiza, Menorca and Mallorca. Rain falls on only three days on average during August and this, in combination with an expected 13 days in the month when sunshine predominates, offers excellent prospects of seeing totality. In the seven years from 2015 to 2021, satellite imagery taken at the same time and on the same date as the 2026 eclipse shows that broken cloud over these islands would have been problematic for some observers located there in only two of these years.

SKY CHART FOR ICELAND, 2026 *Figure 3.3*
Mercury, Venus and Jupiter should be easily visible before totality, together with fainter Mars during totality. Regulus, Capella and Pollux may also be visible.

LUNAR LIMB PROFILE FOR ICELAND, 2026 *Figure 3.4*
Look for a diamond ring at the start of totality at roughly halfway between the 9 o'clock and 10 o'clock positions, and a smaller diamond ring at roughly the 2 o'clock position at the end of totality.

SKY CHART FOR CENTRAL SPAIN, 2026 *Figure 3.5*
Venus should be easily visible before totality, together with Jupiter extremely low in the west-northwest. Regulus, Arcturus and Spica may also be visible during totality.

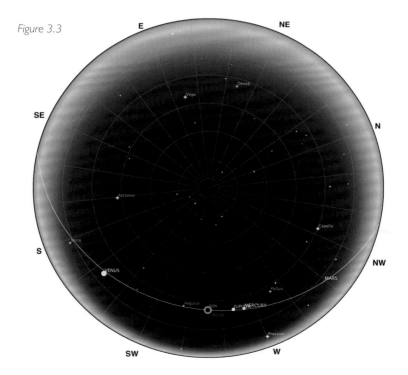

Figure 3.3

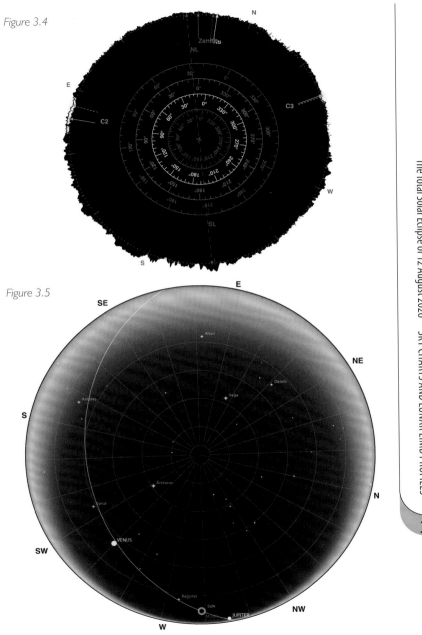

Figure 3.4

Figure 3.5

LUNAR LIMB PROFILE FOR CENTRAL SPAIN, 2026 *Figure 3.6*

Look for multiple Baily's beads, which will transform into a very large diamond ring at the start of totality at roughly the 9 o'clock position, and a large diamond ring at roughly the 4 o'clock position at the end of totality.

SKY CHART FOR THE BALEARIC ISLANDS, 2026 *Figure 3.7*

The chart shows that Venus should be easily visible before totality. Regulus, Arcturus and Spica may also be visible during totality. The eclipse being so low on the horizon should also appear very large. The extended part of the corona will also not be as visible because of the large amount of atmosphere between the viewer and the horizon.

LUNAR LIMB PROFILE FOR THE BALEARIC ISLANDS, 2026 *Figure 3.8*

Look for multiple Baily's beads, which will manifest themselves as a very large diamond ring at the start of totality at roughly the 10 o'clock position, and a large diamond ring at roughly the 4 o'clock position at the end of totality.

Figure 3.6

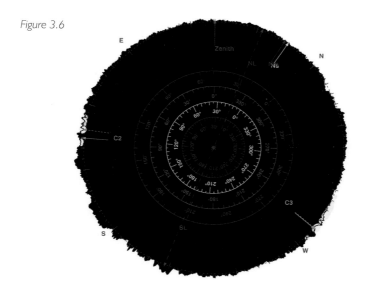

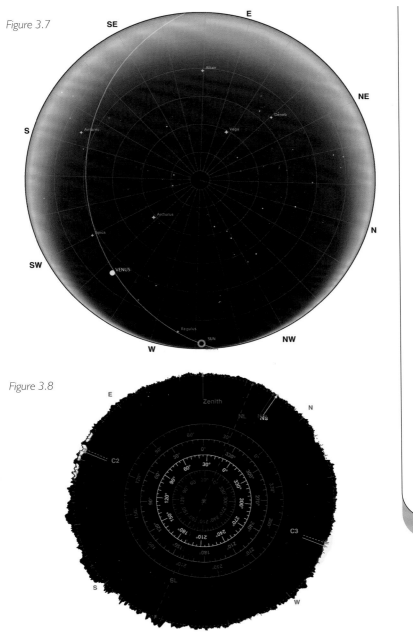

Figure 3.7

Figure 3.8

Spain is fortunate to have three central eclipses cross it during 2026–28. For details of the 2026 total eclipse, see from page 41; the 2027 total eclipse is covered from page 54, and the 2028 annular eclipse from page 79. The best time to watch annular eclipses is during sunrise or sunset, and the Balearic Islands offer this opportunity for 2028.

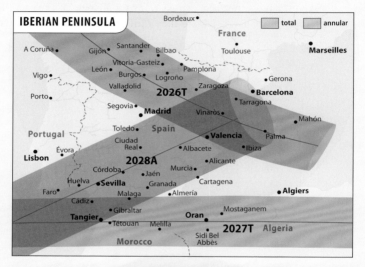

The total solar eclipse of 12 August 2026 passes over the Cantabrian Mountains, in northern Spain.

The total eclipse of 2026 is the forty-eighth member of 72 eclipses in Saros series 126. All eclipses with an even number occur at the Moon's descending node and the Moon moves northwards with each member in the family. The series started on 10 March 1179. It begins with eight partials, 28 annulars, three hybrids, ten totals and ends with 23 more partials. The series ends on 3 May 2459.

The map below shows the 2026 eclipse track together with the previous and next eclipses in the Saros. Notice they shift around the globe, moving west by 8 hours each time.

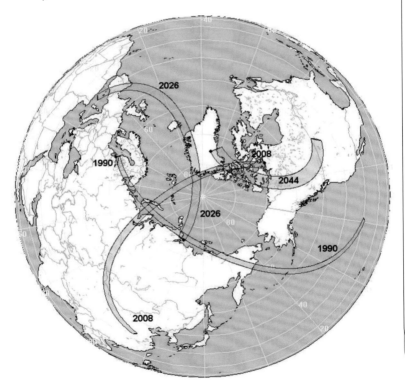

*Diamond ring at the start of totality (approx. 10.10UT (GMT),
during the total solar eclipse of 2015, viewed in Svalbard.*

4

The Total Solar Eclipse of 2 August 2027

WHERE IT GOES *Figure 4.1*

The path of the eclipse starts in the Atlantic Ocean before entering the extreme southwest tip of Spain and Gibraltar. It then crosses Morocco,

Figure 4.1

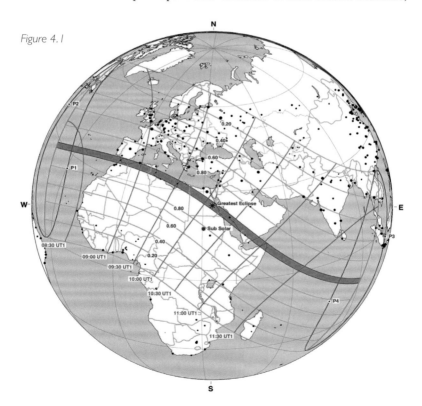

Tunisia, Libya, Egypt, Saudi Arabia, Yemen and Somaliland ending in the Indian Ocean south of the Maldives.

For a detailed map of the track, see w xjubier.free.fr/tse2027map. A map showing the intersection of the 2026 and 2027 totals plus the 2028 annular eclipse crossing Europe can be found on page 50.

CLIMATOLOGY

The total solar eclipse of 2027 is the longest total eclipse for the rest of the century and will pass over some of the sunniest countries in the world. Despite these very sunny climates, some locations do come with an increased risk that local, or global meteorological conditions might 'conspire' against you on eclipse day.

SOUTHERN SPAIN AND NORTHERN MOROCCO The Moon's umbral shadow first makes landfall across southern Spain and northern Morocco. High sunshine amounts are experienced in these regions in August and offer an excellent chance of seeing the eclipse.

However, during settled conditions when a warm, moist wind blows in from the eastern Mediterranean, the Levanter, this can give rise to periods of poor visibility with fog and low cloud particularly for coastal locations surrounding the Alboran Sea in the western Mediterranean. The well-known Banner-cloud that forms over Gibraltar's famous rock is a feature of this moist easterly wind.

Statistics show that there are four days on average when these poorer conditions prevail. This moist air can also affect large parts of the north African coastline, and a little way inland, allowing fog and low stratus cloud to sometimes form overnight; if this fog and cloud do occur on the day of the eclipse, they will more than likely disperse well before the event.

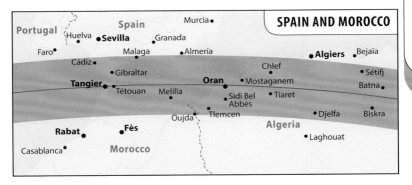

Time of greatest eclipse 10:07:50 GMT
Maximum duration of totality 6m 23s
Location of greatest eclipse 25°30.2'N, 33°11.0'E
Altitude of Sun at greatest eclipse 82°
Maximum width of path of totality 257.7km
Saros Series 136, member 38 of 71
Magnitude 1.07903
Obscuration 1.16430
Gamma 0.14209

ALGERIA, TUNISIA AND LIBYA The eclipse shadow passes over the Tell Atlas mountains in Algeria and eastern Tunisia, which roughly follow the northern coastline. Although sunshine totals are very high in this region, at times these high elevations help generate thunderstorms resulting in great swathes of cloud extending right across the mountains to the coast.

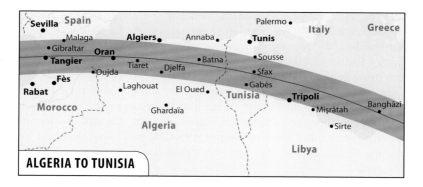

ALGERIA TO TUNISIA

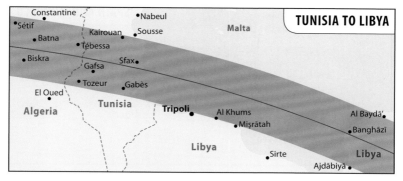

TUNISIA TO LIBYA

It's not until you get to the Gulf of Gabes (Tunisia) that you mostly escape the effects of these thundery episodes.

South of Sfax (Tunisia), totality's centre line again moves back out into the Mediterranean Sea for a time before making contact with Benghazi (Libya). Here, statistics show that the Sun shines for 88% of the time possible with 21 clear days and no predominantly cloudy days.

EGYPT The shadow moves away from the coast and heads into the interior of Egypt. How fortunate it is that the location of maximum duration of

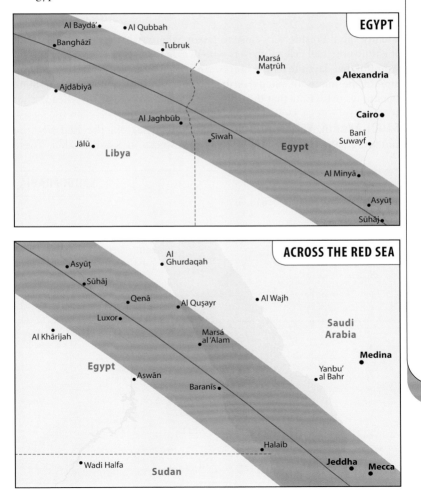

Figure 4.2

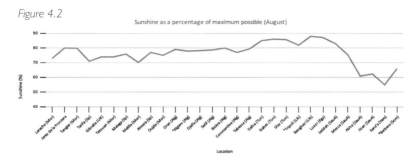

Sunshine as a percentage of maximum possible (August)

totality should reside very close to Luxor, where the Sun shines for 87% of the time possible. Eclipse chasers who decide to observe from here must be mindful that daytime temperatures can exceed 40°C, so adequate provision should me made for plenty of drinking water and shade from the Sun's harsh rays. Also note that the eclipse will be almost exactly overhead.

SAUDI ARABIA, YEMEN AND SOMALIA Moving across the Red Sea the umbral shadow passes over Jeddah and Mecca in Saudi Arabia. These

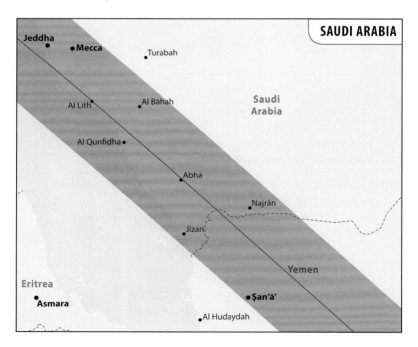

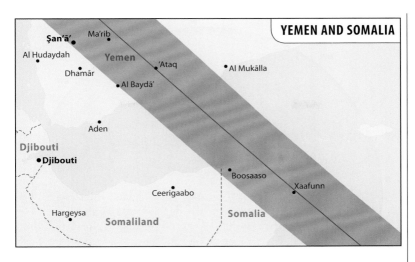

YEMEN AND SOMALIA

locations are both very sunny and hot (see Figure 4.2), but they are not totally immune to the occasional appearance of cloud, which can be generated by nearby mountains or from the influence of the southwest monsoon way to the south. In the southern parts of the country lie the high mountains of the Asir region. Here, the sunshine potential is greatly limited by the monsoon. On average rain is expected to fall on 11 days across the high mountainous areas such as Abha, reducing to four days on the coast, at Jizan.

This theme of limited sunshine continues into Yemen, but in eastern Somalia on the Horn of Africa, the orientation that the coastline presents to the southwest winds of the monsoon allows for coastal divergence and consequently atmospheric subsidence and sunnier conditions. However, this divergence also produces upwelling of cooler water and a greater chance for the formation of mist and fog.

SKY CHARTS AND LUNAR LIMB PROFILES

THE SKY AT TOTALITY OVER SPAIN, 2027 *Figure 4.3*
Mercury and Venus should be visible before totality, with Jupiter evident during totality plus Saturn on the opposite side of the sky. Stars that may be visible include Pollux, Procyon, Sirius, Betelgeuse and Capella.

LUNAR LIMB PROFILE OVER SPAIN, 2027 *Figure 4.4*
Baily's beads will appear at roughly the 9 o'clock position and transition into a very large diamond ring at the start of totality. The end of totality produces a similarly large diamond ring at 4 o'clock.

4

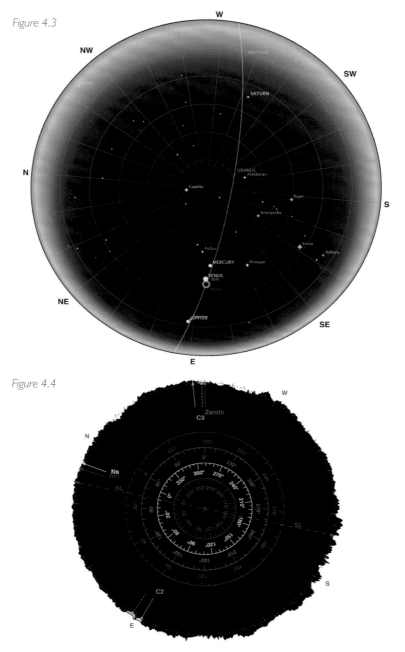

Figure 4.3

Figure 4.4

Figure 4.5

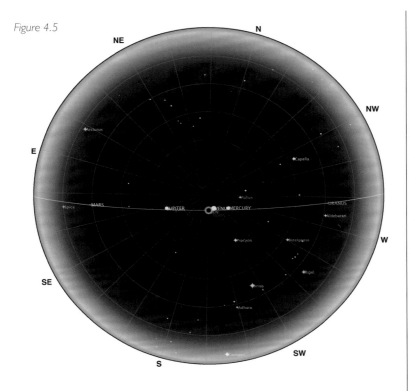

THE SKY AT TOTALITY OVER EGYPT, 2027 *Figure 4.5*

With the Sun almost at the zenith you may be advised to lie back in a reclining chair to view Mercury and Venus before totality, with Jupiter and Mars evident during totality. Stars that may be visible include Pollux, Procyon, Sirius, Betelgeuse, Aldebaran, Arcturus and Capella.

LUNAR LIMB PROFILE OVER EGYPT, 2027 *Figure 4.6*

Baily's beads followed by a small diamond ring will appear at roughly the 8 o'clock position. The end of totality produces a very large diamond ring at 3 o'clock.

THE SKY AT TOTALITY OVER THE HORN OF AFRICA, 2027 *Figure 4.7*

Mercury and Venus should be visible before totality, with Jupiter and Mars evident during totality. Stars that may be visible include Pollux, Procyon, Sirius, Betelgeuse, Arcturus and Spica.

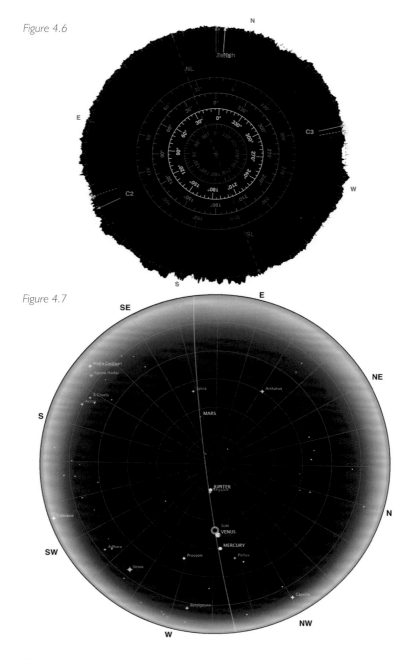

Figure 4.6

Figure 4.7

LUNAR LIMB PROFILE OVER THE HORN OF AFRICA, 2027 *Figure 4.8*

A diamond ring will appear at roughly the 11 o'clock position at the start of totality. The end of totality produces a large diamond ring at 5 o'clock.

Figure 4.8

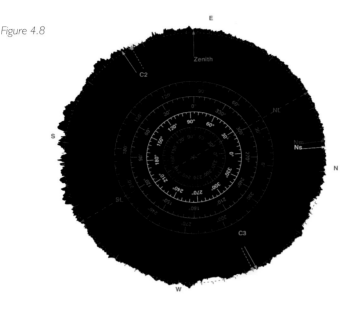

THE SAROS

The total eclipse of 2027 is the thirty-eighth member of 71 eclipses in Saros series 136. All eclipses with an even number occur at the Moon's descending node and the Moon moves northwards with each member in the family. The series started on 14 June 1360. It begins with eight partials,

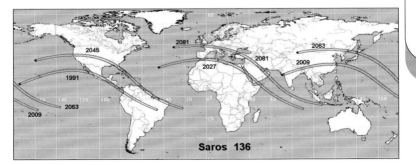

Saros 136

Location	1st	2nd	Mid	3rd	4th	Alt	Dur/%
EUROPE							
Lisbon	07:40:38	–	08:44:40	–	09:54:51	34°	93%
Gibraltar	07:41:01	08:45:28	08:47:40	08:49:54	10:01:02	38°	4m 27s
Seville	07:41:24	–	08:47:24	–	09:59:51	38°	98%
Isla de Tarifa	07:41:45	08:45:00	08:47:19	08:49:39	10:00:36	38°	4m 40s
Malaga	07:41:59	08:48:01	08:48:57	08:49:53	10:02:28	39°	1m 53s
Madrid	07:45:22	–	08:51:08	–	10:02:37	40°	88%
London	08:03:33	–	09:00:04	–	09:59:23	40°	52%
Edinburgh	08:11:05	–	09:00:22	–	09:51:35	37°	38%
MOROCCO							
Rabat	07:38:58	–	08:45:12	–	09:58:32	37°	98%
Tangier	07:40:28	08:44:34	08:46:59	08:49:24	10:00:15	38°	4m 50s
ALGERIA							
Oran	07:44:19	08:50:56	08:53:29	08:56:03	10:09:14	43°	5m 6s
S Algiers	07:48:08	08:58:05	08:58:51	08:59:37	10:15:37	47°	1m 36s
N Algiers	07:48:15	–	08:58:57	–	10:15:38	47°	99.6%
TUNISIA							
Tunis	07:56:05	–	09:09:55	–	10:28:32	55°	97%
Sfax	07:56:17	09:08:40	09:11:29	09:14:20	10:31:38	56°	5m 42s
LIBYA							
Tripoli	07:59:35	–	09:16:43	–	10:38:19	60°	99.94%
Mitiga Airport	07:59:42	09:16:23	09:16:52	09:17:22	10:38:28	60°	57s
Benghazi	08:10:35	09:27:47	09:30:51	09:33:56	10:53:13	68°	6m 6s
EGYPT							
Siwa Oasis	08:22:07	09:42:22	09:45:12	09:48:03	11:07:56	75°	5m 41s
Cairo	08:33:11	–	09:55:59	–	11:16:25	78°	95%
Luxor Centreline	08:40:05	10:01:50	10:05:01	10:08:13	11:26:15	82°	6m 21s
Luxor	08:40:06	10:01:57	10:05:07	10:08:17	11:26:24	82°	6m 19s
Hurghada	08:41:01	–	10:05:05	–	11:25:23	80°	98%
Aswan	08:42:41	–	10:08:17	–	11:29:43	83°	99.4%
SAUDI ARABIA							
Medina	08:57:00	–	10:20:34	–	11:38:18	76°	94
Jeddah	09:00:20	10:22:13	10:25:15	10:28:16	11:43:42	76°	6m 5s
YEMEN							
Sana'a	09:21:42	10:44:05	10:45:11	10:46:15	12:00:27	67°	2m 10s
Al Irqah	09:31:35	10:50:31	10:53:20	10:56:08	12:06:32	62°	5m 37s

Location	1st	2nd	Mid	3rd	4th	Alt	Dur/%
SOMALIA							
Bosaso	09:39:34	10:58:11	11:00:06	11:02:01	12:11:53	58°	3m 50s
Xaafuun	09:44:39	11:01:12	11:03:56	11:06:39	12:14:30	55°	5m 26s
INDIAN OCEAN							
Maldives	10:33:39	–	11:37:26	–	12:34:48	22°	89%

six annulars, six hybrids, 44 totals and ends with seven more partials. The series ends on 30 July 2622.

The map on page 63 shows the 2027 eclipse track together with the previous and next eclipses in the Saros. Notice they shift around the globe, moving west by 8 hours each time.

In this Saros the 2081 eclipse is total in the Channel Islands. Three Saroses later the eclipse of 2135 crosses the UK followed by another in 2189.

5

Annular and Partial Solar Eclipses 2023–2028

INTRODUCTION

Annular solar eclipses are completely different to total solar eclipses. The Sun is never completely covered by the Moon, so darkness never descends on the surroundings and therefore totality cannot be experienced. Also they should only be observed with the use of solar filters or projection.

During a total eclipse there is almost no noticeable darkening until the Sun is 99% covered. During annular eclipses there is always an amount of Sun visible; indeed it can be up to 10%, which is in effect almost normal daylight. As such, hundreds of years ago when eclipses were not predicted, an annular eclipse could have happened without anyone being aware.

The longer the duration of an annular eclipse, the harder it is to see as there is still a significant portion of the Sun showing. In long-duration annular eclipses the Moon is much smaller than the Sun and hence there is very little reduction in light. Because you get extended-duration Baily's beads as the Moon skims the edge of the Sun in an annular eclipse, it is best to observe them at either the northern or southern limit, rather than on

ANNULAR AND PARTIAL SOLAR ECLIPSES 2023–2028

14 October 2023 Annular
2 October 2024 Annular
29 March 2025 Partial
21 September 2025 Partial

17 February 2026 Annular
6 February 2027 Annular
26 January 2028 Annular

A map showing the 2028 annular eclipse crossing Europe and the intersection of the 2026 and 2027 total eclipses is on page 50.

◀ *Partial solar eclipse seen in Penzberg, Upper Bavaria, Germany (4 January 2011).*

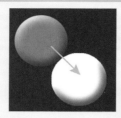

First contact: The moment when the edge of the Moon first encroaches on the Sun.

Second contact: The Moon is now completely inside the Sun.

Mid-eclipse: The Moon doesn't cover the Sun completely.

Third contact: The Moon is just about to leave the Sun.

Fourth contact: The end of the eclipse.

the centre line. With a camera it is even possible to photograph the eclipse without filters and see prominences, provided a very high shutter speed and low ISO are used. The photograph below – taken during the 2005 annular

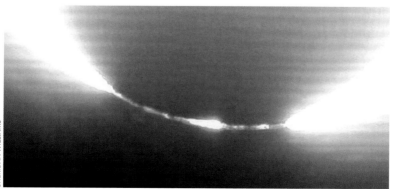

eclipse from Moraira, Spain – is an enlargement from a hugely over-exposed photograph illustrating this effect.

During a **partial solar eclipse**, the Sun, Moon and Earth are not aligned in a perfect straight line, and the Moon casts only the outer part of its shadow, the penumbra, on Earth. From our perspective, this looks like the Moon has taken a bite out of the Sun. The eclipsed area is known as the eclipse magnitude.

THE ANNULAR SOLAR ECLIPSE OF 14 OCTOBER 2023

WHERE IT GOES *Figure 5.1*
The path of the Moon's antumbral shadow starts in the Pacific Ocean, crosses the USA, Mexico, Central America, Colombia and Brazil ending in the mid-Atlantic. For a detailed map of the track, see
w xjubier.free.fr/ase2023map.

Figure 5.1

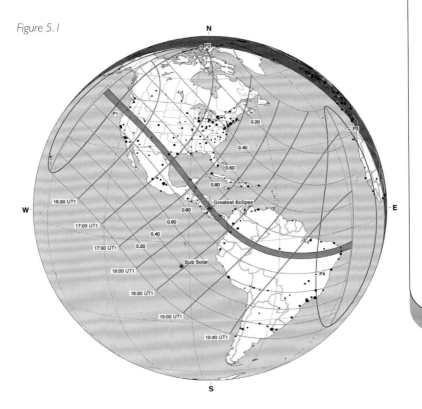

Location	1st	2nd	Mid	3rd	4th	Alt	Dur/%
USA							
Eugene, OR	15:05:27	16:16:57	16:18:54	16:20:50	17:39:46	18°	3m 42s
Richfield, UT	15:09:01	16:26:25	16:28:43	16:31:02	17:56:53	29°	4m 27s
Durango, CO	15:11:48	–	16:34:02	–	18:04:46	33°	94%
Albuquerque, NM	15:13:14	16:34:32	16:36:57	16:39:21	18:09:26	36°	4m 41s
Santa Fe, NM	15:13:34	16:35:59	16:37:24	16:38:48	18:09:50	36°	2m 32s
San Antonio, TX	15:23:49	16:52:05	16:54:16	16:56:27	18:33:00	47°	4m 6s
Corpus Christi, TX	15:26:27	16:55:45	16:58:16	17:00:48	18:38:11	49°	4m 52s
MEXICO							
Merida	15:45:05	–	17:24:15	–	19:08:41	60°	94%
Campeche	15:45:27	17:22:25	17:24:42	17:26:59	19:09:25	61°	4m 19s
Chetumal	15:51:00	17:29:46	17:31:57	17:34:08	19:17:10	63°	4m 2s
BELIZE							
Belize City	15:52:53	17:31:44	17:34:19	17:36:54	19:19:45	64°	5m 0s
HONDURAS							
La Ceiba	15:58:20	17:38:31	17:41:08	17:43:44	19:26:49	66°	5m 2s
NICARAGUA							
Bluefields	16:11:08	17:53:49	17:56:26	17:59:04	19:41:52	68°	5m 4s
COSTA RICA							
Limon	16:16:56	18:02:41	18:03:00	18:03:19	19:48:02	69°	0m 1s
PANAMA							
Santiago	16:25:14	18:10:11	18:12:17	18:14:22	19:56:18	67°	3m 55s
Panama City	16:26:03	–	18:13:18	–	19:56:54	66°	94%
COLOMBIA							
Cali	16:45:35	18:31:35	18:33:25	18:35:14	20:13:45	62°	3m 25s
BRAZIL							
Natal	18:29:28	19:43:54	19:45:41	19:47:29	20:14*	6°	3m 23s
Recife	18:31:57	–	19:47:17	–	20:14*	6°	92%

Annular solar eclipse rising behind the Barnegat Lighthouse, New Jersey, USA (10 June 2021). ▶

THE ANNULAR SOLAR ECLIPSE OF 2 OCTOBER 2024

WHERE IT GOES *Figure 5.2*

The path of the Moon's antumbral shadow spends most of its time in the Pacific Ocean, passing into southern Chile and Argentina before ending in the south Atlantic Ocean. For a detailed map of the track, see **w** xjubier.free.fr/ase2024map.

Figure 5.2

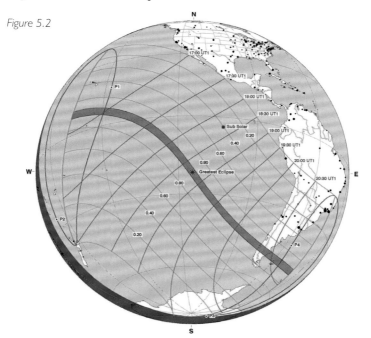

LOCAL ANNULAR ECLIPSE CIRCUMSTANCES, OCTOBER 2024

Location	1st	2nd	Mid	3rd	4th	Alt	Dur/%
Kiribati	–	16:17*	17:00:51	–	18:18:29	10°	81%
Hunga Roa Is	17:23:45	19:04:08	19:07:11	19:10:14	20:52:20	67°	5m 45s
Tortel, Chile	18:56:43	20:19:58	20:23:09	20:26:19	21:41:45	26°	6m 11s
ARGENTINA							
Puerto San Julian	19:03:52	20:24:22	20:27:03	20:29:44	21:42:44	21°	5m 13s
Stanley, Falkland Islands	19:12:57	–	20:30:55	–	21:42:13	14°	90%

*Sunrise

2024 ANNULAR ECLIPSE FACTS

Time of greatest eclipse 18:46:13 GMT
Maximum duration of annularity 7m 25s
Location of greatest eclipse 21°57.2'S, 114°30.5'W
Altitude of Sun at greatest eclipse 69°
Maximum width of path 266km
Saros Series 144, member 16 of 70
Magnitude 0.9326
Obscuration 0.8698
Gamma -0.3509

THE PARTIAL SOLAR ECLIPSE OF 29 MARCH 2025

WHERE IT CAN BE SEEN *Figure 5.3*

The partial eclipse is visible in the north Atlantic from the extreme west of Canada, Greenland, Iceland, western Europe and the northwest of Africa. The magnitude of the eclipse can be seen in the chart. Local circumstances for major cities are given in the table on page 74. For a detailed map of the track, see w xjubier.free.fr/pse2025-03map.

MARCH 2025 PARTIAL ECLIPSE FACTS

Time of greatest eclipse 10:48:36 GMT
Saros Series 149, member 21 of 71
Magnitude 0.9376
Obscuration 0.9306
Gamma 1.0453

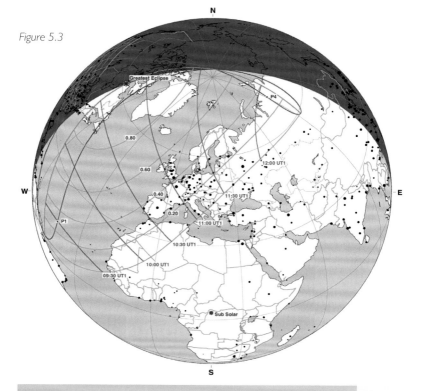

Figure 5.3

PARTIAL SOLAR ECLIPSE CIRCUMSTANCES, MARCH 2025

Location	Start	Max	Rise	End	Alt	Magnitude
New York	09:23:42*	10:12:50*	10:44	11:04:48	0°	27%
Ottawa	09:30:41*	10:20:38*	10:49	11:13:25	0°	22%
Bar Harbor, Maine	09:26:05*	10:17:20*	10:20	11:11:40	0°	82%
St John's, Newfoundland	09:27:41	10:22:52		11:21:38	11°	85%
Reykjavik	10:05:42	11:05:29		12:07:06	24°	74%
Dublin	10:01:30	11:00:18		12:00:53	37°	51%
London	10:07:16	11:03:20		12:00:43	40°	42%
Edinburgh	10:08:34	11:07:00		12:06:48	36°	51%
Lerwick	10:15:59	11:14:24		12:13:49	32°	53%
Marrakesh	09:31:32	10:18:06		11:06:55	47°	26%
Madrid	09:48:41	10:40:14		11:33:48	47°	32%
Paris	10:08:40	11:01:55		11:56:20	43°	35%
Berlin	10:32:18	11:19:34		12:07:03	41°	26%
Oslo	10:30:11	11:24:35		12:19:11	34°	41%
Moscow	11:25:10	11:49:18		12:13:13	31°	6%

* Sun rises after start of eclipse

WHERE IT CAN BE SEEN *Figure 5.4*

The eclipse spends most of its time in the Pacific Ocean, passing into southern Chile and Argentina before ending in the south Atlantic Ocean. For a detailed map of the track, see w xjubier.free.fr/pse2025-09map.

Figure 5.4

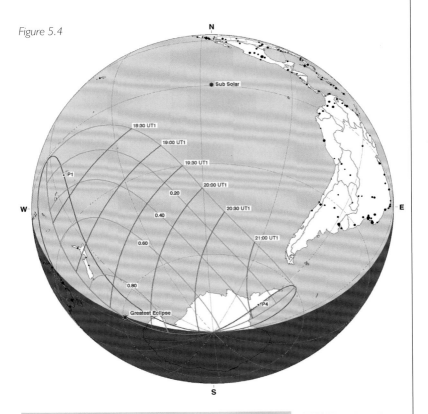

SEPTEMBER 2025 PARTIAL ECLIPSE FACTS

Time of greatest eclipse 19:43:04 GMT
Saros Series 154, member 7 of 71
Magnitude 0.85504
Obscuration 0.7969
Gamma -1.0651

Location	Start	Max	Rise	End	Alt	Magnitude
Auckland	17:52:12	18:55:06	18:12	20:04:34	8°	70%
Tauranga	17:53:21	18:57:03	18:06	20:07:24	9°	70%
Wellington	17:59:24	19:03:55	18:11	20:14:54	9°	74%
Christchurch	18:03:28	19:07:58	18:20	20:18:41	8°	77%
Dunedin	18:07:52	19:12:20	18:28	20:22:44	7°	79%
Invercargill	18:09:12	19:13:09	18:37	20:22:51	5°	80%

THE ANNULAR SOLAR ECLIPSE OF 17 FEBRUARY 2026

WHERE IT GOES *Figure 5.5*

This eclipse is included for the sake of completeness as undoubtedly fewer people will travel so far for an annular eclipse. For a detailed map of the track, see w xjubier.free.fr/ase2026map.

Figure 5.5

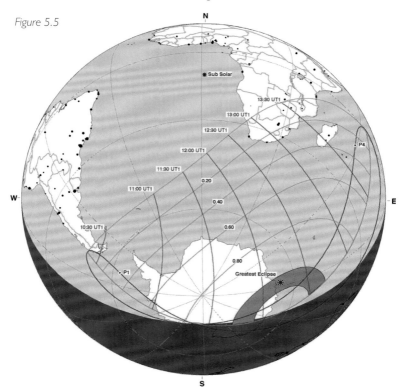

Location	1st	2nd	Mid	3rd	4th	Alt	Dur/%
Ushuaia, Argentina	10:04:32	–	10:31:04	–	10:58:16	7°	9%
Mirny, Antarctica	11:05:56	12:07:13	12:08:09	12:09:06	13:07:45	10°	1m 48s
Durban, South Africa	12:10:04	–	13:05:48	–	13:57:21	46°	27%
Taolagnaro, Madagascar	12:25:18	–	13:24:16	–	14:18:08	28°	41%
Reunion	12:36:30	–	13:32:33	–	14:23:54	18°	43%

2026 ANNULAR ECLIPSE FACTS

Time of greatest eclipse 12:13:06 GMT
Maximum duration of annularity 2m 20s
Location of greatest eclipse 64°43.1'S, 086°44.4'E
Altitude of Sun at greatest eclipse 12°
Maximum width of path 616km
Saros Series 121, member 61 of 71
Magnitude 0.9630
Obscuration 0.9274
Gamma -0.9743

THE ANNULAR SOLAR ECLIPSE OF 6 FEBRUARY 2027

WHERE IT GOES *Figure 5.6*

The eclipse starts south of Easter Island in the Pacific Ocean, crosses southern Chile and Argentina skimming Uruguay and Brazil, crossing the

2027 ANNULAR ECLIPSE FACTS

Time of greatest eclipse 16:00:48 GMT
Maximum duration of annularity 7m 53s
Location of greatest eclipse 31°18.2'S, 048°28.0'W
Altitude of Sun at greatest eclipse 73°
Maximum width of path 281.0km
Saros Series 131, member 51 of 70
Magnitude 0.9281
Obscuration 0.8614
Gamma -0.2952

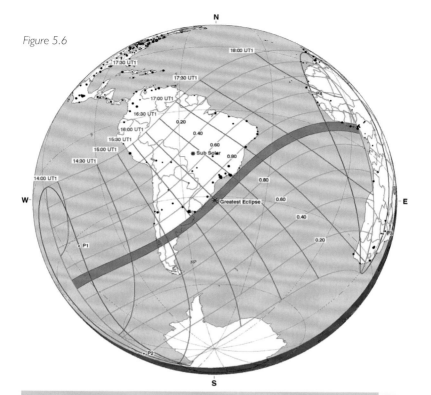

Figure 5.6

LOCAL ANNULAR ECLIPSE CIRCUMSTANCES, FEBRUARY 2027

Location	1st	2nd	Mid	3rd	4th	Alt	Dur/%
ARGENTINA							
Mar del Plata	13:45:13	15:24:15	15:28:04	15:31:53	17:10:02	66°	7m 30s
URUGUAY							
Montevideo	13:50:24	–	15:36:08	–	17:19:07	70°	92%
Punta del Este	13:52:47	15:35:56	15:38:38	15:41:21	17:21:02	70°	5m 0s
BRAZIL							
Rio de Janeiro	14:40:33	–	16:29:34	–	18:04:59	69°	90%
GHANA							
Accra	16:41:50	17:49:37	17:52:22	17:55:07	18:12*	4°	5m 21s
TOGO							
Lome	16:42:56	17:49:59	17:52:38	17:55:17	18:06*	2°	5m 10s
BENIN							
Porto Lovo	16:43:45	17:49:59	17:52:40	17:55:21	18:00*	1°	5m 13s
NIGERIA							
Lagos	16:44:01	17:49:46	17:52:33	17:55:19	17:57	0°	5m 24s

*Sun sets before end of eclipse

Atlantic Ocean, touching land again in the Cote d'Ivoire, Ghana, Togo, Benin and Nigeria. For a detailed map of the track, see w xjubier.free.fr/ase2027map.

THE ANNULAR SOLAR ECLIPSE OF 26 JANUARY 2028

WHERE IT GOES *Figure 5.7*
The eclipse starts in the Galápagos Islands crossing Ecuador, Peru, southern Colombia, Brazil, southern Suriname and French Guiana. Next it crosses the Atlantic Ocean into Madeira, Portugal, Spain, Gibraltar and Morocco, ending at sunset in the Balearic Islands. For a detailed map of the eclipse track, see w xjubier.free.fr/ase2028map.

CLIMATOLOGY Annularity makes first landfall over the Galápagos Islands (Ecuador) in the western Pacific – made famous by the 19th-century naturalist Charles Darwin. The time of the eclipse coincides with the

Figure 5.7

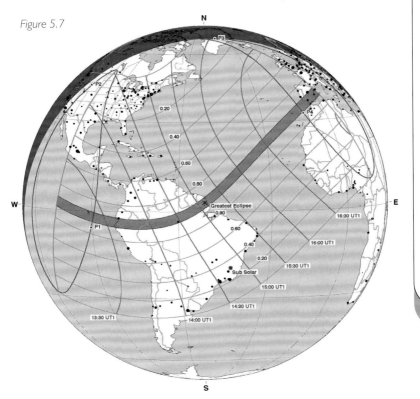

Time of greatest eclipse 15:08:58 GMT
Maximum duration of annularity 10m 27s
Location of greatest eclipse 02°57.6'N, 051°33.6'W
Altitude of Sun at greatest eclipse 70°
Maximum width of path 323.0km
Saros Series 141, member 24 of 70
Magnitude 0.9208
Obscuration 0.8479
Gamma 0.3901

islands' wet season, but the average rainfall total for January at Puerto Baquerizo Moreno (Island San Christóbal) is a relatively meagre 50mm. This rain comes mostly in the form of afternoon showers which fall much later than the time of the eclipse although this diurnal pattern is lost during those years when El Niño prevails.

The path of annularity heads towards continental South America, passing over the Gulf of Guayaquil and making landfall on the coasts of Ecuador and northern Peru. During January Ecuador's climate is tropical, being both hot and humid, with heavy thundery downpours. But right on the coast at locations such as Salinas (Ecuador) and Talara (Peru) the influence of the cool waters of the Humboldt ocean current reduces localised convective cloud, allowing drier and sunnier conditions than found inland such as at Guayaquil. However, these same cool waters do form large patches of slowly drifting low marine stratus and stratocumulus cloud. In the lowlands, away from the coast, sunshine increases southwards, within the eclipse track, from 27% of the maximum possible at Guayaquil (Ecuador) to 58% in the tropical and arid climate of Piura in Peru. During times of El Niño, rainfall increases dramatically with flooding highly likely in coastal and lowland regions.

Moving eastwards across Ecuador into the highlands you find the city Cuenca (2,560m) which is situated within a valley surrounded by high mountains. This location gives rise to a multidirectional rain-shadow and consequently rainfall totals are much lower and sunshine higher than many lowland locations to the west such as Guayaquil. Continuing eastwards the Andes mountains fall away and are replaced by the low-lying hot and humid interior of northern Peru, extreme southern Colombia and the tropical rainforests of Brazil's Amazon Basin. This region receives large amounts of rainfall in January with an average of

255mm, falling on 17 days, at Iquitos (Peru) in the west, to 295mm, falling on 18 days, at Macapá (Brazil), close to the mouths of the Amazon River, in the east. Despite these high totals, January's heavy rainfall is diurnal

LOCAL ANNULAR ECLIPSE CIRCUMSTANCES, JANUARY 2028

Location	1st	2nd	Mid	3rd	4th	Alt	Dur/%
GALÁPAGOS							
Isla Isabella	12:07:30	13:20:36	13:24:00	13:27:25	14:58:15	16°	6m 39s
Machala	12:10:03	13:33:00	13:57:00	13:41:01	15:26:15	30°	7m52s
Quito	12:12:57	–	13:42:04	–	15:34:29	32°	86%
PERU							
Iquitos	12:15:00	13:45:39	13:50:00	13:54:23	15:48:08	40°	8m 33s
BRAZIL							
Manaus	12:34:53	14:14:24	14:28:04	14:31:43	16:33:22	60°	7m 8s
FRENCH GUIANA							
Vila Velha	13:04:58	15:04:05	15:09:19	15:14:31	17:08:43	67°	10m 17s
MADEIRA							
Funchal	15:19:53	16:46:38	16:50:07	16:53:35	18:08:10	18°	6m 46s
PORTUGAL							
Faro	15:32:24	16:51:38	16:55:09	16:58:39	18:08*	9°	6m 55s
SPAIN & GIBRALTAR							
Seville	15:34:23	16:52:14	16:55:52	16:59:29	17:41*	7°	7m 9s
Gibraltar	15:35:09	16:54:23	16:56:26	16:58:28	17:42*	7°	3m 56s
Malaga	15:36:02	16:54:08	16:56:38	16:59:07	17:36*	6°	4m 53s
Madrid	15:36:08	–	16:55:34	–	17:25*	0°	90%
Valencia	15:39:01	16:53:21	16:56:52	17:00:23	17:14*	2°	6m 55s
Barcelona	15:40:10	16:53:20	16:56:30	16:59:39	17:00*	0°	6m 11s
Ibiza	15:40:34	16:55:16	16:57:29	16:59:42	17:08*	1°	4m 20s
Palma	15:41:14	16:54:45	16:57:25	17:00:05	17:02*	0°	5m 13s
MOROCCO							
Tangier	15:34:41	16:54:56	16:56:19	16:57:43	17:44*	8°	2m 34s
OTHER PLACES							
Penzance	15:33:40	–	16:48:31	Sunset 17:06*		1°	66%
London	15:35:53	–	16:48:35	Sunset 16:38*		0°	60%
Paris	15:37:45	–	16:51:28	Sunset 16:38*		0°	70%

*Sun sets before end of eclipse

in nature and comes in the form of heavy convective showers, often growing into afternoon thunderstorms, driven by the heat of the day. Overnight, cloud from the preceding day's downpours decays, meaning that the following day is likely to dawn clear, or maybe with scattered cloud, before the heating cycle begins again. The time of annularity, in the west (Guayaquil), comes about 2 hours after sunrise, meaning that cloud growth will still be limited; however, further east (Macapá) the eclipse comes later in the morning (local time) and with increasing cloud cover. Occasionally, disturbances in the general easterly flow give rise to periods of more general, longer-lasting rain, that doesn't follow the expected convective diurnal cycle.

The path of annular eclipse just clips the extreme southern part of Suriname before leaving the coasts of French Guiana and Brazil's State of Amapa. Heading northeast across the Atlantic, the Moon's antumbral shadow passes over the small Portuguese island of Madeira. January can be stormy for this island group with days of persistent cloud. On Funchal the average monthly rainfall in January is 64mm, falling on six days, and daily average sunshine amounts to 5–6 hours.

By the time the annular phase of the eclipse reaches the mainland of southern Portugal, the southern half of Spain and the extreme north of Morocco, the sun is low in the southwestern sky. Sunshine amounts are relatively high across this area; however, Atlantic low-pressure areas with their cloud and rain do occasionally invade this region during periods when blocking high-pressure areas are centred over the UK or further east over Europe, deflecting weather systems south-eastwards. As stated in other eclipses narratives, even when the sky is generally clear, low-elevation eclipse observations mean there is always a chance that the eclipse could be obscured by patches of scattered cloud along your line of sight, appearing to merge together close to the horizon.

6

Planning, Preparation and Photography

HEALTH AND SAFETY

Each of the eclipses traverses different climatic and geographical regions, so you should consult country guides and specialist health information for advice on safety, climate, health and immunisation before you travel. A full list of current travel clinic websites worldwide is available on w istm.org. For other journey preparation information, consult w travelhealthpro.org. uk (UK) or w wwwnc.cdc.gov/travel (USA). All advice found online should be used in conjunction with expert advice received prior to or during travel.

Some first-time visitors to countries not on the normal tourist trail may initially experience a degree of culture shock, unless cocooned inside first-class hotels and whisked through the countryside in luxury coaches. Make sure you plan your trip carefully, research each destination and once there think twice before the eclipses lure you off the beaten track unless you know from experience how to prepare for such ventures, or have an experienced guide with you. In many of the eclipse countries there are regions where simple mistakes like forgetting to carry drinking water, getting stuck in sand or taking the wrong clothes and footwear, can lead to disaster. The usual means of recovery – internet connection, shops, telephones, medical facilities, other vehicles – may simply not be there.

VIEWING ECLIPSES SAFELY It is dangerous to look directly at the Sun at all times except for the brief period of totality, when it is *fully covered* by the Moon. Some eye experts argue that you should never look directly at an eclipse, even during totality, because it is all too easy to make a mistake and cause permanent damage to your eyes – amid the tremendous excitement of the experience, protocol can be forgotten, you may look too soon, or glasses, especially worn by a child, may slip. Astronomers and other experts say sensible adults should easily be able to follow the basic viewing rules to witness one of the great sights of nature.

Looking with the naked eye during the brief period of totality itself is perfectly safe, but you must use eclipse glasses or other means, eg: projection, during the partial phases. To stare at the Sun is to look directly at a huge, thermonuclear explosion spewing out ultraviolet and infrared rays. Evolution has honed us to find it painful to look at the ultraviolet rays – if your gaze accidentally settles on the Sun, discomfort quickly forces you to look away. But it is the infrared, rather than the ultraviolet, that does most of the damage. Infrared is heat, and heat cooks the delicate tracery of blood vessels at the back of the eye. If your retina is damaged, you will become blind at the spot where images are focused, making it hard to read or recognise people's faces. Nothing in medical science can reverse this. When the Sun is partially eclipsed or annular, there may be insufficient ultraviolet rays to make it painful to gaze at it – but there will definitely be enough infrared to damage the eye. Even at 99% eclipse the remaining rays from the Sun are extremely dangerous.

The one time when it is safe to watch the eclipse is during totality – the short period when the Sun is totally covered by the Moon. The Sun is then only as bright as the full Moon and cannot cause damage. For the rest of the eclipse you *must not* look with the naked eye. It takes mighty screening power to block the Sun's rays and the only suitable materials are specially made, and generally coated with a thin layer of metal such as silver or aluminium. Often the material is aluminised mylar or Baader AstroSolar Safety film; welders' goggles with a rating of 14 or higher are also suitable. Nothing else is safe, and that includes sunglasses, smoked glass, compact discs, exposed black-and-white film, and the sun caps or solar eyepieces provided with binoculars and amateur telescopes (you need special metal-coated filters for these).

For covering your camera lens, telescope or binoculars you will need filters you can cut to shape. You can order eclipse glasses from many companies, but most tour companies will provide them for their clients. If you wish to purchase your own, search the internet for suitable suppliers, checking for ISO certification. (In 2015, the ISO issued a standard for solar viewing material, called ISO 12312-2, written by Dr Ralph Chou. Dr Chou is renowned within the eclipse community as the world's leading expert on eclipse eye safety. He wrote the ISO standard for eclipse glasses that has been adopted for certifying safe eclipse glasses and filter material, and his laboratory certifies all new products as being safe for direct solar observation. Most eclipse glasses manufacturers conform to this standard. You can read his paper at w eclipse.aas.org/sites/eclipse.aas.org/files/Chou-Solar-Eclipse-Eye-Safety-AAS-2016.pdf. Information on suppliers can be found on page 101.

Test your glasses beforehand against a bright light source such as a reading lamp. If there are bright spots visible, throw the glasses away. Ideally equip yourself with several pairs in case one becomes damaged – and keep them where they cannot get scratched.

Watch the stages of the eclipse through the approved glasses but, even with this protection, do not stare for longer than half a minute (damage from gazing directly at the Sun has been reported with less than a minute's viewing). When the diamond ring appears you can remove the glasses, replacing them when it returns at the end of totality.

A completely safe way to view an eclipse, which avoids looking directly at the Sun, is through a pinhole projector. This can be made by piercing a hole in a piece of card with a knitting needle, or similarly narrow implement, and viewing, on another piece of card, the light that shines through it. DO NOT look through the hole!

Another way of observing the partial phases is by use of a Solarscope. This device comes as a flat-packed box with a metal enclosed lens. It allows several people at once to look at the Sun's projected image. Once constructed it becomes a stiff cardboard structure darkened on the inside. It has a lens at the end of the metal fitment plus a small adjustable convex mirror that reflects the Sun's image back into the box. Solarscopes are available from various suppliers found online.

LOCAL EYE SAFETY As nature conducts its own experiment in the sky – the biggest laboratory of all – many people will be unaware of the dangers of eclipse viewing and risk damaging their eyesight for life. As a tourist, your presence may encourage local people to stay outside and watch the drama, unprotected. Alternatively, you can be a source of information (and viewing glasses), so that those who would otherwise have stared long and hard at the eclipse without the necessary preparation may learn how to view it safely.

Several groups are working to help local people make the most of the eclipses while safeguarding their sight. Below are some ways you could help:

- Take several viewers with you.
- Even if you have only one viewer with you, lend it to those without. You will find you want to look at the partial eclipse only briefly, once every 10 minutes or so. One viewer can therefore serve several people.
- Find out the official eye safety advice (if any) in the country you are visiting and try to work with this rather than confusing people by contradicting it.
- Remember the golden rule: do not look at any stage of the partial eclipse except through specially made viewers, projection or a solarscope.

WHAT TO TAKE

Eclipse-viewing glasses are vital. If you plan to take photographs you need to give some time and thought to equipment (see below). You may also want to take binoculars with the appropriate solar filters. A method of recording sound might seem like a strange thing to bring to a visual spectacle but the responses from the people around you will be tremendously vocal, and you can add your own descriptions as you watch. (Alternatively, some people pre-record minute-by-minute instructions about what they should look for and when, and then play the recording back when the eclipse starts.) Thin sheets of cardboard from which to make a pinhole camera may be useful; they can also be used to make holes to project tiny crescent Suns onto the ground. In addition, a tripod, a seat or rug, a torch, Sun protection, mosquito repellent (for when the insects emerge in the twilight) and a water bottle can come in handy.

PHOTOGRAPHY

I well remember that I wished I had not encumbered myself with apparatus, and I mentally registered a vow, that, if a future opportunity ever presented itself for my observing a total eclipse, I would… devote myself to that full enjoyment of the spectacle which can only be obtained by a mere gazer.
>   Warren De La Rue, on the total solar eclipse in Spain of 18 July 1860

If this is your first total solar eclipse, consider first whether you want to be distracted during the precious few minutes of totality by the need to focus cameras and adjust settings. You may absorb the moment better if you forget about photography and ask fellow travellers for copies (in my experience they will be flattered and only too pleased to provide these), or buy professional photographs later – there will be many of them and they will be superb.

Fred Espenak has pages on his website dedicated to eclipse photography: w mreclipse.com.

COMPACT CAMERAS OR SMARTPHONES These are ideal for capturing the surroundings during totality, but unless they have a very large zoom, the eclipse itself will be but a tiny dot in the distance. Instead, I recommend that you turn off the flash (or better still cover it with opaque tape). This is very important for two reasons: first, you will annoy other eclipse

◀1 *Camera and telescopes ready to photograph the 2012 annular solar eclipse in Monument Valley, Utah, USA.* **2**, **3** & **4** *Eclipse glasses are essential to watch the partial phases of a solar eclipse safely.*

watchers; and, second, you want to record the reducing light levels and the flash will negate this. Instead, mount the device on a tripod (because the long exposures will induce shake and give blurred images); you won't need a filter as you are not pointing the camera directly at the Sun. Point the camera horizontally (not up at the Sun) and make sure the memory can hold at least 30 exposures). If the device will allow you to set the exposure and shutter speed manually, set it to 1/50s and ISO400 using the widest aperture possible. Now take shots as follows:

Time to second contact	Frequency of photo
5 minutes	every 1 minute
1 minute	every 20 seconds
Totality	every 10 seconds

Don't try to fiddle with exposure settings during totality, but do turn the camera around on its tripod to change the view. If you can overlap the images during totality by 20%, there are many panorama-stitching programs available to create a truly memorable scene.

SLR, MIRRORLESS AND MEDIUM FORMAT CAMERAS For totality itself you need to use a camera that is capable of using a lens with an effective focal length of at least 300mm, and the text below gives advice on using these. You do not need a telescope.

Filter Don't forget that gazing through your camera viewfinder at the Sun is dangerous – the rays will fry your retina, as well as the electronics inside your camera. You must have a neutral density filter that cuts out light and heat by a factor of 100,000. This translates to an ND5.0 filter. You must use it even when the Sun is 99% covered, removing it just before Baily's beads appear.

Baader AstroSolar filter sheets (page 101) come with instructions on how to make your own camera and binocular filters.

It is also a good idea to cover the whole camera and tripod with a sheet or towel between exposures during the 1 hour from first to second contact as the camera can overheat in the Sun.

Lenses Before choosing the lens to attach to your camera it is important to realise that there is a correlation between the size of the camera's imaging sensor and the image size using the same lens. Many digital SLRs produce an image 1.6 times larger with the same focal length lens when used on a full-frame digital camera. So a 200mm lens attached to a digital camera will produce an image equivalent to 1.6 × 200 = 320mm lens attached to a full-frame camera (making the Sun appear 1.8mm actual size).

In order to get a 2mm image of the eclipsed Sun (ie: just to discern it as a disc) you will need at least a 250mm lens. However, for eclipse close-ups you will need a lens with a focal length of 500mm or more. A 500mm lens will yield a 4.5mm diameter image. Lenses larger than 500mm are unlikely to be able to fit in the whole of the Sun's corona at maximum exposure. Do not use longer focal lengths unless you want to image the Sun's disc without the corona. To work out the size of the image on the resulting photograph, divide the focal length of the lens (in millimetres) by 110 for a full frame camera. Note that you can also buy lens multipliers, and a 2× multiplier will make your 250mm lens into a 500mm lens. Also note that the eclipses will drift across the viewfinder quickly with long focal length lenses.

If you want to be very professional, you can also buy small portable refracting telescopes that are ideally suited to eclipse work. I use a William Optics 66mm (350mm focal length) telescope, but there are many other suitable models. If you wish to follow this route, you are advised to approach a reputable telescope specialist as it is critical for the optics to be of good quality. Please think twice before buying a telescope from anyone else. Bear in mind that additional equipment like a small telescope could well incur extra luggage penalties when travelling.

Specialist telescope and camera shops will sell adaptors for connecting cameras to telescopes. Do not use the eyepiece filter that is furnished with some small telescopes – it is not safe.

Tripod, cable release and sequence control The final essential piece of equipment is a tripod. Lightweight tripods frequently cause blurred eclipse photographs but are better than no tripod at all. The tripod legs should not be extended more than halfway. Try adjusting the height so that you can easily reach the camera controls while sitting on a chair, thereby minimising the vibrations you transmit. You can further decrease vibrations by suspending a weight under the tripod – rocks or sand in a sack. Also use a cable (or remote control) release to stop camera shake when you press the shutter. Some digital cameras have cable releases that are programmable, and many cameras can be controlled by a smart phone or laptop so you can pre-determine every shot and sequence (however, without a driven camera mount, the Sun will drift across the camera's field of view quite quickly, and you will need to keep adjusting the position of the camera). You may also find a right-angled eyepiece adaptor or adjustable screen useful as many eclipse locations have the Sun high in the sky. If you want to automate the whole set-up you will need a driven camera mount. There are many options for this and an online search for 'Star tracker camera mounts' will bring up a huge range of choices. Again, bear in mind the extra weight in your luggage.

How to image using an SLR camera There are two key factors that will determine the success of a close-up eclipse image: focus and camera shake.

Focus is most difficult to cope with and requires practice. If you are using a standard lens, it is not simply a matter of setting the lens to ∞ (infinity). Practise at home before you go, and spend some time during the partial phases fine-tuning the focus, since this can change as the camera and lens heat up. Many cameras are now 'mirrorless' or have 'live view' whereby you can look at the image on the screen at large magnifications.

Details of how to minimise camera shake are detailed below, but there is also the option of using 'mirror lock-up' ('live view' will do this for you) and ensuring the camera is set to the fastest possible shutter speed. Some cameras are mirrorless and solve the problem that way.

Total solar eclipses are among the most tolerant things to photograph. Provided you have a large enough lens and a tripod, every shot you take will show something of interest.

Simple approach Put the camera on a tripod and preferably attach a cable release or remote control, set the camera to automatic exposure and make sure the flash is off. The ISO setting is not crucial, but I have found when using a tripod that ISO >800 will allow you to use the fastest shutter speed to minimise camera shake. You must also turn off the auto-focus as the low light conditions usually prevent this from working. Spend some time during the partial phases finding the precise focus. You will probably over-expose the inner corona and prominences, but you will produce reasonable pictures. This approach has the great advantage of being simple and leaving you time to watch the eclipse properly.

For those first-timers who are serious about wanting to photograph the event, please bear in mind that a total eclipse is intensely emotional, and an awe-inspiring spectacle. You will almost certainly be overcome by the sight and will probably not be able to achieve what you set out to do. Nevertheless the golden rule is preparation. First, decide what kind of pictures you want – evocative scenes of the Sun fringed by trees and

1 A safe way to show the partial phases of an eclipse is to project them on to a negative film sheet. 2 Annular solar eclipse viewing party at Marina Barrage, Singapore (26 December 2019). ▶

You've heard of trainspotters – well, there are people whose sole aim in life is to go to the ends of the Earth (literally) to see a total solar eclipse. These people are given names such as eclipse chasers, ecliptomaniacs, eclipsoholics and umbraphiles. We (I admit that I class myself as one) take total solar eclipses very seriously. For example, I use total solar eclipses to plan my holidays. In 1988 I went to Sumatra, in 1990 to Finland, 1991 the Baja Peninsula in Mexico, 1994 Peru, 1997 Siberia, 1998 Antigua, 1999 Cornwall in England, 2001 Madagascar, 2006 Libya, 2008 the Gobi Desert, 2009 Wuhan in China, 2010 Tahiti, 2012 Australia, 2015 Faroe Islands, 2016 Sulewesi, 2017 USA and 2019 Argentina. What I am really waiting for is another totality of over 6 minutes which will happen in Luxor, Egypt, in 2027 (details in this guide). The next total eclipse that will happen with an amazing backdrop is on 22 July 2028 from Sydney Harbour Bridge, Australia.

Umbraphiles will go to any lengths to find the best location to maximise the chance of seeing totality; their eclipse-viewing locations are posted at w eclipse-chaser-log.com/eclipse-log.

Glenn Schneider is one such umbraphile renowned for his exploits, which include a hike up a mountain on Atka Island in the Aleutians to see the 1990 eclipse. However, his most extraordinary endeavour was on 3 October 1986. A solar eclipse was due to happen but unfortunately it was only to be annular from the surface of the Earth – and an annular eclipse is no good for an umbraphile as the Moon is not big enough to cover the Sun completely. It just so happened that the Moon's umbral shadow (the only place in which a total solar eclipse can be seen) stopped 12km above the Earth's surface. A group of nine umbraphiles led by Glenn arranged to charter a Cessna Citation aircraft to fly up into the arctic air near Greenland and rendezvous with the Moon's tiny shadow, which was probably only a few hundred metres across. If their calculations were correct, and the pilot could fly accurately, they would witness a fleeting glimpse of totality, maybe only one or two seconds, but long enough to notch up another total eclipse. This feat was successful and, as soon as totality ended, Glenn and his team looked down at the cloud tops and saw a cigar-shaped shadow of the Moon racing off into the distance, secure in the knowledge that they were the only living creatures to see this totality. To read more about this and his other adventures, visit w eclipse-chaser-log.com/eclipse-log/291.

The fanatical umbraphile will be armed with software that can predict local circumstances, as well as a detailed chart of the Moon's valleys and mountains referred to as 'lunar limb profiles' (LLP). Using these (included in the chapters for total solar eclipses only), it is often found that maximum totality can be

ERICA RUTH NEUBAUER/S

Totality of over 6 minutes will happen over Luxor, Egypt, in 2027. Pictured here, sphinx statues at the temple complex of Karnak.

some way away from the centre line, allowing a valuable second or two to be added to the viewing log.

The best locations are frequented by thousands of umbraphiles, which in turn attracts the media, local dignitaries and others. This prevents umbraphiles from making exaggerated claims about their times under totality.

If you succeed in seeing totality, I assure you that you will want to see it again – and then I'm afraid *you* will have become an umbraphile.

The author's log can be found at: w *eclipse-chaser-log.com/eclipse-log/99.*

people, or a close-up of the flaring prominences? Try out all new equipment and rehearse the procedure at home (the Sun will conveniently be there to practise on). This way everything has a higher chance of working perfectly at the crucial moment. To simulate totality, wait until half an hour after the Sun has set and practise changing the shutter speed from shortest to longest without a torch.

If you are photographing a partial eclipse, or the partial phases of a total and annular eclipse, you should calculate the shutter speed and aperture appropriate to your solar filter and telephoto lens by setting up the apparatus on a sunny day at home. Do make sure that you are using the highest resolution setting your camera has (usually known as RAW setting). Again, practise all this at home on the Sun several weeks before the eclipse. If your camera, has a removable memory card, consider buying another one for use during the eclipse.

More complex approach For this, set the camera to manual and make sure that RAW (or RAW + JPEG) mode is set. While still at home practise by pointing the camera at the Sun (using your solar filter) and set the aperture to f/8. Take a frame for every shutter speed from 1/4000s (or whatever is your shortest) to 1/2s, record the settings and view the results to see which works best. Bear in mind that if the sky is hazy at eclipse time, you will need longer exposures to compensate. This approach allows you to practise getting the precise focus too.

Prior to totality Before the eclipse begins, set the camera to maximum resolution, fit the filter, set the focus to infinity (some lenses focus beyond infinity and you will have to experiment to find precise focus – it's worth spending some time doing this). Switch off auto-focus, switch to manual, turn off the flash and set the camera to f/8 and the shutter speed you determined by experimenting at home (f/8 is a good compromise allowing the use of fastest shutter speeds, minimising camera shake and yielding the best results across the whole frame). Do not hunt through the sky for the Sun without the filter fitted. Bear in mind that the partial phases last roughly an hour, so the Sun will move across the field of view and the camera will need regular positional adjustment. Don't forget the cable release, if you have one.

Depending on how many exposures you have available, take one exposure at first contact, then one every 5 minutes (probably 12 exposures). Keep noting which way the Sun travels in the viewfinder so that, when totality arrives, you have the Sun positioned so that it remains in the viewfinder for the duration of totality minutes (this will be difficult, if not impossible, for the long-duration 2027 total eclipse; page 54). If you are using a focal length of more than 400mm and totality lasts more than 2 minutes, you will almost certainly have to move the camera at least once during totality itself.

It is recommended that 1 minute prior to the second contact (and the resulting diamond ring), you locate the Sun in the viewfinder, with the filter still on, in the position determined in the last paragraph (so in 2 minutes it will not drift out of the other side), then, *and only then*, carefully remove the filter so as not to move the camera.

During totality The difficult bit is determining the shutter speed for the build-up to totality. You don't want to have to change settings more often than you need. It will be dark and you may not even be able to see the dial or screen! So, assuming you have set the ISO to 800 and the aperture to f/8, I suggest that you set the shutter speed to 1/2000s for the diamond ring and Baily's beads. Using the cable release, fire off shots every 2 seconds (unless you have a very stable tripod when you can image more frequently even using motor-drive). When all beads are gone and totality is upon you, it becomes so dark that you will need immediately to increase the exposure successively in steps from 1/1000s to 4s depending on what phenomenon you want to capture.

Each eclipse phenomenon has a different brightness value so exposures will vary according to what you want to photograph. For example, to capture the solar prominences and chromosphere with the suggested f ratio and ISO, you need an exposure time of about 1/1000s. But to capture the extreme outer corona, the exposure time should be as long as 4s; this should also show details on the Moon illuminated by earthshine. The solution is to bracket your exposures. Using the technique to change just the shutter speed you have practised at home in the dark, proceed as follows once the diamond ring has gone and totality is in progress:

- Start with a shutter speed of 1/1000s exposure, wait 5 seconds
- Change the shutter to 1/500s exposure, wait 5 seconds
- Change the shutter to 1/250s exposure, wait 5 seconds
- Keep doing this until the shutter speed is 4s

This will have occupied just over 2 minutes (if you are at the eclipse limits or near the end of the track, you will have to reduce the frequency as appropriate as totality is much shorter). Whatever you do, once you reach 4s exposure, then pause and drink in the spectacle. You can always resume a little while later (if you are not too awe-struck). If you can, set the exposure back to 1/2000s, move the eclipse back to the centre of the viewfinder and wait for the final diamond ring. When it occurs, keep pressing the shutter (or use the camera's multi-exposure setting if it has one) as the diamond ring could last 4–5 seconds.

Alternatively, simply reverse the sequence as totality ends to capture prominences on the other limb of the Sun and the final Baily's beads and diamond ring. Once totality is finished, do not forget to replace the filter, and consider covering the camera as the Sun gets more exposed.

At sea At sea, eclipse photography is constrained by the movement of the ship. It is unlikely that you could use a focal length of more than 500mm because of this. You must also contend with vibration from the ship's engines, wind across the deck, and other passengers' footsteps. On a ship it is best to use a high ISO setting in order to be able to use the fastest shutter speeds possible. Notice the range of motion of the ship and attempt to snap the picture when it reaches one extreme.

USING A CAMCORDER OR DSLR WITH VIDEO FACILITY TO CAPTURE THE ECLIPSE These can be used in two modes:

- First, experiment at home to find the zoom setting that allows the Sun plus at least two solar diameters either side to fit in the frame. This is because the corona can extend some way beyond the Sun. To take the eclipse itself, mount the camera, with a filter (page 101), on a tripod and set the camera to automatic. Remove the filter during totality (or slightly before) and continue. If the camcorder has a manual mode, try increasing the exposure during totality to bring out the fine structure of the corona. Make sure any built-in illuminating light is disabled. This can all be practised at home on the un-eclipsed Sun many weeks in advance.
- Second, to film the observing site and the people around, including their reaction to totality, put the camera on a tripod and set it to automatic with the zoom at its widest setting. Rotate the camera before and during totality. Make sure any built-in illuminating light is disabled. If the camera has a manual mode, fix the exposure about 10 minutes before totality and do not vary it; this way you will show the increasing level of darkness as the eclipse progresses.

Fred Espenak has advice on recording eclipses on his website w mreclipse.com.

7

Conclusion

Hopefully you will be reading this before you have seen these eclipses, in which case the conclusion for you has not yet happened, and what a conclusion it will be!

If you are reading this after the eclipse, you should now realise what all the fuss was about, and why people travel to the ends of the Earth to be submerged in totality, and why only totality will do. Once you've seen a solar eclipse you can add yourself to the list of umbraphiles at w eclipse-chaser-log.com/eclipse-log, and log the time you spend under the Moon's shadow. Marvel at the fact that some people have travelled to more than 30 total solar eclipses. Once you have joined the club and seen more than one total eclipse, you will almost certainly be hooked and want to see more.

Total eclipses are infrequent enough to allow time to save sufficient money to travel, and they offer the opportunity of experiencing places that you might otherwise not consider. If you choose your location wisely, there is always so much else to see at your chosen destination.

If you've never seen an eclipse from a cruise ship, then do not rule it out. Long-exposure photography is more difficult, but the camaraderie you'll experience makes up for this. As almost all eclipses have large parts of their track over the oceans, you often have no alternative but to join a cruise. Another option, especially if the weather prospects are poor, is to consider a flight. These are becoming increasingly popular; however, they can be very expensive and you sacrifice experiences seen on the ground for an almost certain chance of a view unobstructed by clouds.

The total eclipse map on page 127 provides a good overview of the eclipse tracks until 2040, and you can use the dates and information on page 102 to help plan your holiday.

I wish you success and clear skies in your eclipse chasing, and I'm sure you will enjoy meeting like-minded people in the most unlikely places.

PAUL COLEMAN

8

Further Information

BOOKS

Espenak, Fred and Anderson, Jay *Eclipse Bulletin: Total Solar Eclipse of 2024 April 08* Available at w eclipsewise.com/pubs/EB2024.html. This guide gives technical details about the next total solar eclipse through North America. It is organised in six sections: (1) Eclipse predictions and umbral path; (2) Local circumstances for the eclipse; (3) Detailed maps of the umbral path; (4) Weather prospects for the eclipse; (5) Observing the eclipse; and (6) Eclipse resources.

Folley, Tom and Zaczek, Ian *The Book of the Sun* Courage Books. To understand more of the folklore and mythology surrounding the Sun.

Guillermier, Pierre et al *Total Eclipses* Springer, 1999. Heavily weighted towards the science, this book has a less personal tone and is for those who really want to get into the history of the scientific study of eclipses.

Hoskin, Michael, ed. *The Cambridge Illustrated History of Astronomy* Cambridge University Press, 1997. If humanity's past responses to eclipses has whetted your appetite for the history of astronomy, try this book.

Littman, Mark, Willcox, Ken and Espenak, Fred *Totality: Eclipses of the Sun* Oxford University Press, 1999. A well-written exposition of the science, culture and history of eclipses, set alight by the authors' enthusiasm.

McEvoy, J P *Eclipse: The Science and History of Nature's Most Spectacular Phenomenon* Fourth Estate, 1999. The most sugared scientific pill I have found, it is strong on the astronomical and cultural history of eclipses and easy to read.

Pasachoff, Jay and Menzel, Donald *A Field Guide to the Stars and Planets* Houghton Mifflin, 1992. Find out more about the heavens fleetingly revealed during the eclipse.

Stephenson, Richard F *Historical Eclipses and Earth's Rotation* Cambridge University Press, 1997. Full of interesting quotations about eclipses as well as following – and solving – an intriguing scientific mystery using the eclipse calculations of the Chaldeans.

◀ *Total solar eclipse from Antalya, Turkey (2006).*

Williams, Sheridan *UK Solar Eclipses from Year 1 to 3,000* Clock Tower Press, 1996. Contains an anthology of 3,000 years of solar eclipses – an intriguing catalogue of the stories attached to Britain's eclipses including those that have started wars, accompanied new kings and symbolised death. Also included are all the eclipse facts you could possibly wish for, a detailed look at eclipse mechanics and discussion of the practicalities of viewing eclipses. Published by the author of this guide. For a copy, email e sw@clock-tower.com with the subject UKSE (£9.95 plus p&p: £1 in UK, £2 Europe/Eire, £3 elsewhere).

Zeiler, Michael and Bakich, Michael E *Atlas of Solar Eclipses – 2020 to 2045.* Available at: w greatamericaneclipse.com. The website also contains links to ephemera such as eclipse glasses etc.

Zeiler, Michael and Bakich, Michael E *Field Guide to the 2023 and 2024 Solar Eclipses.* Also available at: w greatamericaneclipse.com. A book with unbelievably detailed maps of the tracks of the two eclipses by ace cartographers.

WEBSITES

GENERAL

w **earthview.com** For tutorials.

w **eclipsechaser.com** A potpourri of interesting eclipse information.

w **greatamericaneclipse.com** Information, maps, photos and videos related to all types of solar eclipses not just in the US, but around the world.

w **mreclipse.com** Fred Espenak's website has a plethora of publications that can be ordered, including *Eclipse Bulletin: Total Solar Eclipse of 2024 April 08* co-authored with Jay Anderson (page 99).

w **web.williams.edu/Astronomy/IAU_eclipses** For the uses of eclipses to science.

w **xjubier.free.fr/en/index_en.html** Xavier Jubier's amazing eclipse site with Google Earth maps and useful links.

WEATHER

w **eclipsophile.com** Jay Anderson is well known as the weather expert for eclipses. His website is the one to access for detailed information on all aspects of climate and weather for upcoming eclipses. Jay says: 'Every provider will be using EUMETSAT's satellite images, so the choice is in the display and its frequency rather than a different point of view.'

w **meteoblue.com** A good website covering worldwide locations. Detailed cloud-cover maps show likely problems that may reduce the probability of success.

w **rammb-slider.cira.colostate.edu** High-resolution colour images of just about anywhere can be viewed on this website. Pick Meteosat 11 to view Spain and zoom in. Lots of wavelengths available.

ECLIPSE SOFTWARE AND APPS

Because the average length of totality is around 2 minutes, it is essential to make the most of this very short period to view and capture as much as possible. You can use your smartphone, camera or computer to assist with this; and there are various things that software and apps can help with:

- Giving you a running commentary as the eclipse progresses
- Providing precise timings
- Controlling your camera
- Simulation

Visit the American Astronomical Society solar eclipse website (w eclipse. aas.org/resources/apps-software) for a list of the ever-increasing number of packages available. There is also a dedicated website, w eclipse2024.org/eclipse-simulator, that simulates both the 2023 annular and 2024 total solar eclipses that pass through North America. This is free for online use only (no download) – but it is embeddable via iframe.

Of particular interest, Gordon Telepun has the Solar Eclipse Timer app, which is available for download for both iOS and Android. This is a 'talking' eclipse timer with verbal announcements for all the important events including audible countdowns to the contact times and max eclipse. It is free to download and play with, but data sets cost US$1.99.

Gordon has also released *Eclipse Day – 2024 and More! How to enjoy, observe and photograph a total solar eclipse*. This is a multimedia ebook, available for both Apple and Android tablets, providing a fresh concept in eclipse books for eclipse preparation and education. The chapters, 28 of them, are explanations of the important segments of eclipse day from arriving at the observing site, up to leaving at the end of the eclipse. More information can be found at w solareclipsetimer.com.

Stellarium (w stellarium.org) is a free open-source planetarium for your computer. It shows a realistic sky in 3D, just like you would see with the naked eye, binoculars or a telescope. It can also show eclipses far into the past or the future.

EQUIPMENT AND ACCESSORIES SUPPLIERS

For covering your camera lens, telescope or binoculars you will need filters that can be cut to shape. These are available from the stockists listed

Further Information   EQUIPMENT AND ACCESSORIES SUPPLIERS

8

below; or simply search online using the terms 'Baader solar filters' or 'Baader AstroSolar' to find suppliers. Eclipse-viewing glasses are also available via w greatamericaneclipse.com/eclipse-viewing.

UK

Harrison Telescopes
w harrisontelescopes.co.uk
Tring Astronomy Centre
w tringastro.co.uk
Widescreen Centre
w widescreen-centre.co.uk

USA

For a list of stockists in the USA, visit w eclipse.gsfc.nasa.gov/SEpubs/19990811/ text/filter-sources.html

ASTRONOMY TOUR OPERATORS

UK

Ancient World Tours   w ancient.co.uk
Astro Trails   w astro-trails.com
Cruise.co.uk   w cruise.co.uk
Explore Worldwide   w explore.co.uk/experiences/eclipse-trips
Naturetrek   w naturetrek.co.uk/tour-focus/eclipse-and-astronomy
On the Go Tours   w onthegotours.com
Sirius travel   w siriustravel.com

USA

A Classic Tour   w aclassictour.com
Black Tomato   w blacktomato.com/experience-types/eclipse-travel
Eclipse Traveler   w eclipsetraveler.com
Exodus Travel   w exodustravels.com/cultural-holidays/eclipse
Ring of Fire Expeditions   w eclipsetours.com
Spears Travel   w spearstravel.com
Travel-Quest   w travelquesttours.com
Tropical Sails   w tropicalsails.com
Tusker Trail   w tusker.com

FUTURE TOTAL AND ANNULAR SOLAR ECLIPSES

22 July 2028 Australia – Total
1 June 2030 North Africa, Siberia, Japan – Annular
25 November 2030 Australia – Total
21 May 2031 South Africa, India, Indonesia – Annular
14 November 2031 Pacific Ocean – Hybrid
30 March 2033 Alaska – Total

9

Eclipse Destinations A–Z

You can find reliable advice on travelling abroad, including the latest information on safety and security, via **w** gov.uk/foreign-travel-advice. Always check before travel. Other useful websites are listed under individual destinations on the following pages.

National and public holidays are given for dates occurring around one month either side of the day of the total eclipse. For Spain, the 2026 and 2027 eclipses occur in the north and south of the country respectively and two sets of public holidays have been produced for the different regions accordingly (page 121).

Population figures for main towns and cities are for the city (not metropolitan) area only, unless otherwise stated.

ALGERIA *Telephone code +213*

Neighbouring countries Morocco, Tunisia, Libya, Mauritania, Mali, Niger
Area 2,381,741km^2
Climate The coastal strip experiences hot dry summers and mild wet winters. The mountains are cold in winter and hot in summer with less rain than the coast. The desert is very hot in summer and mild in winter.
Status Unitary semi-presidential republic
Population 44,700,000; 2021 estimate
Life expectancy 78.1 years for women, 76.2 years for men (WHO, 2020–21)
Capital Algiers (population 4,510,000; 2011)
Other main towns and cities (population; date) Oran (4,130,039 / metropolitan area 6,089,731; 2008/2019), Constantine (464,219; 2008), Sétif (288,461; 2008), Annaba (464,740; 2019), Djelfa (490,248; 2018), Biskra (307,987; 2007)
Economy (major industries) Imports: US$40,430 million (2019–20). Exports US$37,322 million (2021) – petroleum gas, crude petroleum, nitrogen fertilisers. Data from the UN.
GDP US$4,151 per capita; US$187.155 billion (2022 estimate)
Languages Arabic, Tamazight, Algerian Arabic (Darja), French
Religion Sunni Muslim 99%, other, including Christian and Jewish, 1%
Currency Algerian dinar (DZD)
Exchange rate £1 = 165DZD, US$1 = 137DZD, €1 = 146DZD (Feb 2023)

- **Algiers** Kasbah – UNESCO World Heritage Site.
- **Oran** Fort Santa Cruz gives great views of the city.
- **Timgad** (Batna Province) Roman ruins (AD100) – UNESCO World Heritage Site.
- **Biskra** (430km southeast of Algiers) The Gateway to the Sahara.
- **Tipasa** A coastal town whose roman ruins are a UNESCO World Heritage Site.
- **Djémila** (Sétif Province) Roman ruins – UNESCO World Heritage Site.

National airline Air Algérie
Airports Algiers Houari Boumediene, Oran Ahmed Ben Bella, Constantine Mohamed Boudiaf
Time UTC +1 hour (CET)
Electrical voltage 230V (50Hz)
Public holidays Mawlid, The Prophet's Birthday 14–15 August 2027
National tourist board website w tourismalgeria.com
Red tape A visa will be required to visit Algeria, applying through your local Algerian embassy/consulate. Visas are generally issued within 10–20 working days, although the processing time can vary according to the applicant's nationality and the visa type. Useful websites: w embassypages.com/algeria, w visaguide.world/africa/algeria-visa.

CANADA *Telephone code +1*

Neighbouring countries/regions USA, Greenland
Area 9,984,670km²
Climate Inland, the vast continent experiences freezing winters and hot summers. The climate along the Atlantic coast is tempered but colder and much wetter than the pacific seaboard.
Status Federal parliamentary constitutional monarchy
Population 36,991,981 (2021)
Life expectancy 84.1 years for women, 80.4 years for men (WHO, 2020–21)
Capital Ottawa (Ontario) (population 1,017,449; 2021)
Other main towns and cities (city population/metropolitan area) Toronto (2,794,356 / 6,202,225; 2021), Montreal (1,762,949 / 4,291,732; 2021),

1 *Santa Cruz fort of Oran, Algeria.* **2** *Pharaoh Rameses II statue, Luxor Temple, Egypt.* **3** *The Windsor Suspension Bridge affords spectacular views from Gibraltar.* **4** *A boat tour is one of the best ways to view Niagara Falls, Canada.* ▶

ANTON_IVANOV/S

MATIAS PLANAS/S

JUAN ANTONIO ORIHUELA/S

ANJELIKAGR/S

- **Fundy National Park** (New Brunswick Province)
- **Magnetic Hill** (New Brunswick Province) Near Moncton, a famous site of optical illusion where water appears to flow uphill.
- **Toronto** (Ontario Province) 1,815ft CN Tower with glass floors; Royal Ontario Museum.
- **Montreal** (Quebec Province) Parc Jean-Drapean with giant biosphere from 1967 World Fair; Old Port of Montreal; La Ville Souterraine underground city.
- **Niagara Falls** (Ontario Province) On border with New York State.

Vancouver (662,248 / 2,642,825; 2021), Calgary (1,306,784 / 1,481,806; 2021), Edmonton (1,010,899 / 1,418,118; 2021)

Economy (major industries) Imports: US$405,001 million (2019–20). Exports: US$503,726 million (2021) – gold, timber, automotive industry, aircraft industry, vegetable oil. Data from the UN.

GDP US$56,794 per capita; US$2.20 trillion (2022)

Languages English, French

Religion Christian 53.3%, no religion 34.6%, Muslim 4.9%, Hindu 2.3%, Sikh 1.4%

Currency Canadian dollar (CAD)

Exchange rate £1 = 1.62CAD, US$1 = 1.35CAD, €1 = 1.44CAD (Feb 2023)

National airline Air Canada

Airports Toronto Pearson, Vancouver, Montréal-Pierre Elliott Trudeau, Calgary

Time Newfoundland Time: NST (UTC -3hrs 30mins), NDT (UTC -2hrs 30mins); Atlantic Time: AST (UTC -4hrs), ADT (UTC -3hrs)[1]; Eastern Time: EST (UTC-5hrs), EDT (UTC -4hrs)[1]; Central Time: CST (UTC -6hrs), CDT (UTC -5hrs)[1]; Mountain Time: MST (UTC -7hrs), MDT (UTC-6hrs)[1]; Pacific Time: PST (UTC -8hrs), PDT (UTC -7hrs).

[1]Some regions within specified time zone remain on standard time all year-round.

Electrical voltage 120V (60Hz)

Public holidays Good Friday (except Quebec), Easter Monday (Quebec only)

National tourist board website w travel.gc.ca

Red tape Most people will require a visa or an Electronic Travel Authorisation to travel to Canada. Visa processing time may be two to four weeks, or even longer, depending on your exact requirements. Useful websites: w travel.gc.ca (Government of Canada website, which has advice on the visa process); w embassypages.com/canada.

Neighbouring countries Libya, Sudan, Israel

Area 1,010,408km^2

Climate Predominantly a desert climate with hot and dry conditions in summer and mild in winter with the highest rainfall near the Mediterranean coast.

Status Unitary semi-presidential republic

Population 107,770,524 (2022 estimate)

Life expectancy 74.1 years for women, 69.6 years for men (WHO, 2020–21)

Capital Cairo (population 10,100,166 / metropolitan area 21,900,000; 2022)

Other main towns/cities (population) Alexandria (6,050,000; 2022), Giza (9,200,000; 2021), Shubra El Kheima (1,099,354; 2012), Port Said (818,677; 2020), Suez (744,189; 2018)

Economy (major industries) Imports: US$78,657 million (2019–20). Exports: US$40,701 million (2021) – refined petroleum, gold, crude petroleum, nitrogenous fertilisers. Data from the UN.

GDP US$4,176 per capita; US$438.348 billion (2022 estimate)

Languages Arabic

Religion Muslim 90.3%, Christian 9.6%

Currency Egyptian pound (EGP)

Exchange rate £1 = 36.79EGP, US$1 = 30.57EGP, €1 = 32.66EGP (Feb 2023)

National airline Egyptair

Airports Cairo, Hurghada, Luxor, Bord El Arab (Alexandria)

Time UTC +2hrs (Egypt Standard Time, EST)

Electrical voltage 220V (50Hz)

Public holidays Revolution Day (23 July 2027), The Prophet's Birthday (14–15 August 2027).

National tourist board website w egypt.travel

Red tape A visa will be required in most cases to visit Egypt and can be applied for through your local embassy or consulate, or online (e-visa). E-visas are usually processed within five working days, but do allow at least

EGYPT HIGHLIGHTS

- **Alexandria** Site of the Lighthouse of Alexandria, one of the seven wonders of the ancient world.
- **Cairo** Pyramids of Giza and Egyptian Museum of Antiquities.
- **Luxor** The Valley of the Kings; Temple of Karnak.
- **Aswan** Temple of Philae; Aswan's Botanical Garden.
- **Abu Simbel** Great Temple of Ramses II.

9

one week for processing before departure, in the event of delays. Applying through your local embassy/consulate may take at least 10 working days. Useful websites: w embassypages.com/egypt; w visa2egypt.gov.eg (e-visa applications).

GIBRALTAR *Telephone code +350*

Neighbouring countries Spain

Area 6.8km^2

Climate Summers tend to be warm and sunny and winters, mild and wet.

Status British Overseas Territory: devolved representative democratic parliamentary dependency under constitutional monarchy.

Population 34,003; 2020 estimate

Life expectancy 81.7 years for women, 77.1 years for men (WHO, 2021)

Capital Westside (population 26,572; 2012)

Economy (major industries) Imports: US$8,663 million (2019–20). Exports: US$103 million (2021) – banking and finance, ship repair. Data from the UN.

GDP GBP£71,787 per capita; GBP£2.441 million (2021 estimate)

Languages English, Spanish, Llanito

Religion Catholic 72.1%, Protestant 7.7%, Muslim 3.6%, Jewish 2.4%, Hindu 2.0%

Currency Gibraltar pound (GIP)

Exchange rate £1 = 1GIP, US$1 = 0.83GIP, €1 = 0.89GIP (Feb 2023)

National airline Gibraltar Airways

Airports Gibraltar Airport

Time Winter: UTC+1 hour (CET); Summer: UTC+2hrs (CEST)

GIBRALTAR HIGHLIGHTS

- **Upper Rock Nature Reserve** Take the cable car or steps to the top of the rock to see the famous Barbary Apes and fantastic views across the Mediterranean to Morocco.
- **St Michael's Cave** More than 150 caves, including the Cathedral Cave.
- **The Moorish Castle** Originally built in the 8th century.
- **Main Street and Grand Casemates Square** Numerous restaurants, cafés and shops.
- **Gibraltar Skywalk and Windsor Suspension Bridge** A new attraction perched 340m above the Mediterranean. Not for those who are afraid of heights.

Electrical voltage 230V (50Hz)
Public holidays Late Summer Bank Holiday (30 August 2027)
National tourist board website w visitgibraltar.gi
Red tape A visa is not required for Australian, UK, EU or USA citizens to visit Gibraltar. Other nationalities should visit w gibraltarborder.gi to ascertain their visa requirements.

GREENLAND (KALAALLIT NUNAAT) *Telephone code +299*

Neighbouring countries Canada
Area 2,166,086km²
Climate Subarctic with very cold winters. Summers are cool although periods of warmth can occur in the south.
Status Part of the Kingdom of Denmark – devolved government within a parliamentary constitutional monarchy.
Population 56,466 (2022 estimate)
Life expectancy 71 years for women, 71 years for men (WHO, 2020–21)
Capital Nuuk (population 18,800 / metropolitan area 19,023; 2021)
Other main towns and cities (population) Sisimiut (5,582; 2020), Ilulissat (4,670; 2020), Aasiaat (3,069; 2020), Qaqortoq (3,050; 2020), Maniitsoq (2,534; 2020)
Economy (major industries) Imports: US$1,015 million (2019–20). Exports: US$1,366 million (2021) – economy is dependent on Danish subsidies, making up over 50% of government revenue; fishing and fish products make up 90% of Greenland's exports. Data from the UN.
GDP US$53,000 per capita; US$3.0 billion (2019 estimate)
Languages Greenlandic (Kalaallisut), Danish
Religion Protestant 95.5%, Catholic 0.2%, other Christian 0.4%, Inuit spirited belief 0.8%, agnostic/atheist 2.5%, other religion 0.6%
Currency Danish krone (DKK)
Exchange rate £1 = 8.39DKK, US$1 = 6.97DKK, €1 = 7.45DKK (Feb 2023)
National airline Air Greenland
Airports Godthaab/Nuuk

GREENLAND HIGHLIGHTS

- **Greenland National Museum** (Nuuk)
- **Ilulissat Icefjord** UNESCO World Heritage Site.
- **Jakobshavn Glacier** The most productive glacier in the northern hemisphere.
- **Northeast Greenland National Park**

9

Time Most of Greenland operates under a standard time of UTC -3hrs and daylight saving time UTC -2hrs. Danmarkshavn UTC +0hrs; Scoresbysund UTC -1 hour. Qaanaaq UTC -4hrs.

Electrical voltage 230V (50Hz)

Public holidays None during the eclipse period

National tourist board website w visitgreenland.com

Red tape A visa is not required for Australian, UK, EU or USA citizens to visit Greenland for visits of up to 90 days. Other nationalities should visit w nyidanmark.dk or w embassypages.com/greenland to ascertain their visa requirements.

ICELAND *Telephone code +354*

Neighbouring countries None

Area 102,775km^2

Climate Cold in winter but milder, particularly near west and south coasts. Summers are mild with periods of warmth. Sunshine is limited and rainfall is abundant throughout the year.

Status Unitary parliamentary republic

Population 376,248; 2022 estimate

Life expectancy 83.9 years for women, 80.8 years for men (WHO, 2020–21)

Capital Reykjavík (population 131,136 / metropolitan area 233,034; 2020)

Other main towns and cities (population) Kópavogur (37,959; 2020), Hafnarfjörður (29,971; 2020), Reykjanesbær (19,724; 2021), Akureyri (19,219; 2021), Garðabær (16,924; 2020)

Economy (major industries) Imports: US$5,706 million (2019–20). Exports: US$5,976 million (2021) – fish, fish products, aluminium, medicines and ferrosilicon. Data from the UN.

GDP US$73,981 per capita (2022); US$20.8 billion (2020 estimate)

Languages Icelandic (official), English, German

Religion Evangelical Lutheran Christian (60.9%), other Christian (11.5%), no religion (25.2%), Asatru (1.5%)

Currency Icelandic króna (ISK)

Exchange rate £1 = 174.27ISK, US$1 = 144.78ISK, €1 = 154.70ISK (Feb 2023)

National airline Icelandair

Airports Keflavik (near Reykjavík), Akureyri

Time UTC +0hrs year-round

Electrical voltage 230V (50Hz

◀ *1 Eruption of Strokkur Geyser, Iceland. 2 Colourful houses overlook Greenland's Ilulissat Icefjord.*

9

- **Blue Lagoon** (Grindavík)
- **Strokkur Geyser**
- **Gullfoss Waterfall**
- **Mælifell Volcano and Myrdalsjökull Glacier**
- **The Perlan Museum and Planetarium**
- **Vatnshellir Cave** 8,000-year-old lava tube.
- **Snæfellsjökull** National Park and glacier with 700,000-year-old stratovolcano.

Public holidays Commerce Day (3 August 2026)
National tourist board website w visiticeland.com
Red tape A visa is not required for Australian, UK, EU or USA citizens to visit Iceland for visits of up to 90 days. Other nationalities should visit w government.is or w embassypages.com/iceland to ascertain their visa requirements.

LIBYA *Telephone code +218*

Neighbouring countries Algeria, Tunisia, Niger, Chad, Sudan, Egypt
Area 1,759,541km²
Climate The coastal strip experiences a Mediterranean climate with hot summers and mild winters with rainfall mostly confined to the autumn/winter months. The desert region is very hot in summer, although cooler in the mountains, and mild in winter.
Status Unitary provisional unity government
Population 7,054,493 (2022 estimate)
Life expectancy 77.3 years for women, 74.2 years for men (WHO, 2020–21)
Capital Tripoli (population 1,170,000; 2019)
Other main towns and cities (population) Benghazi (807,250 / metropolitan area 1,110,000; 2020), Misrata (386,120; 2011), Bayda (380,000; 2010), Khoms (201,943; 2004), Zawiya (200,000; 2011)
Economy (major industries) Imports: US$13,123 million (2019–20). Exports: US$30,640 million (2021) – crude petroleum, petroleum gas, scrap iron, refined petroleum. Data from the UN.
GDP US$7,197 per capita; US$50.326 billion (2022 estimate)
Languages Libyan Arabic, Berber, Tamasheq, Teda, Italian
Religion Islam (99.7%), other (0.3%)
Currency Libyan dinar (LYD)
Exchange rate £1 = 5.77LYD, US$1 = 4.79LYD, €1 = 5.12LYD (Feb 2023)

- **Leptis Magna** One of the most beautiful cities of the Roman Empire – UNESCO World Heritage Site.
- **Cyrene** Magnificent ruins of ancient Greek and Roman temples and townhouses – UNESCO World Heritage Site.
- **Tobruk** Place of fiercest battles between Allied and Axis forces during World War II.
- **Misrata** City with multicultural architecture, 'sun-kissed' beaches and sand dunes.

National airlines Libyan Airlines, Afriqiyah Airways
Airports Tripoli
Time UTC +2hrs (Eastern European Time, EET)
Electrical voltage 230V (50Hz)
Public holidays The Prophet's Birthday (14–15 August 2027)
National tourist board website w Libya Tourism.com
Red tape Because of the fragile security situation (crime, terrorism, civil unrest, kidnapping and armed conflict) within Libya, the UK Foreign, Commonwealth and Development Office (w gov.uk/foreign-travel-advice/libya) and their Australian, EU and US equivalents are advising against any travel to this country. There is limited consular support.

MEXICO *Telephone code +52*

Neighbouring countries USA, Guatemala, Belize
Area 1,972,550km^2
Climate There are a number of different climate regions across Mexico. From the arid northern mountains and hot northwest desert in summer to the seasonably wet conditions of the central plateau, that contains Mexico City. The warm Caribbean creates a subtropical/tropical climate in the east with copious rainfall and occasional tropical storms in summer. The tropical west is hot all year and very wet in summer.
Status Federal presidential republic
Population 129,150,971; 2022 estimate
Life expectancy 78.9 years for women, 73.1 years for men (WHO, 2020–21)
Capital Mexico City (population 9,209,944 / metropolitan area 21,804,515; 2020)
Other main towns and cities (population) Tijuana (1,810,645; 2020), Ecatepec (1,643,623; 2020), León (1,579,803; 2020), Puebla (1,542,232; 2020), Ciudad Juárez (1,501,551; 2020)

Eclipse Destinations A–Z MEXICO

9

- **Mazatlán** (Sinaloa State) Machado Square, picturesque tree-lined plaza area of old town.
- **Ojuela Suspension Bridge** (Durango State) Much smaller prototype of San Francisco's Golden Gates; Ojuela's abandoned goldmine and town nearby.
- **Torreón** (Coahuila State) Spectacular mountain views (via cable car) of Torreón and distant Durango; Canal de la Perla, subterranean tunnel.
- **Saltillo** (Coahuila State) Museo de Revolucion Mexicana; Parque Ecologico El Chapulin.
- **Garcia Caves** (Nuevo León State) Caverns and caves 30km from Monterrey.
- **Gran Plaza O Macroplaza** (Monterrey City) The macroplaza is the fifth largest in the world.
- **Valladolid** Cenote Zaci, caverns, caves and other geological formations.

Economy (major industries) Imports: US$467,293 million (2019–20). Exports: US$494,595 million (2021) – machinery and transport equipment, steel, electrical equipment, chemicals, food products, petroleum products. Data from the UN.

GDP US$10,950 (2022) per capita; US$1.42 trillion (2022 estimate)

Languages Spanish (de facto)

Religion Catholic 74%, no religion (14.8%), Protestant (4.3%), other (6.9%)

Currency Mexican peso (MXN)

Exchange rate £1 = 22.13MXN, US$1 = 18.38MXN, €1 = 19.64MXN (Feb 2023)

National airline Aeroméxico

Airports Mexico City, Cancún, Guadalajara, Monterrey

Time Zona Sureste Time: ST (UTC -5hrs); Zona Centro Time: ST (UTC -6hrs), DT (UTC -5hrs); Zona Pacífico Time: ST (UTC -7hrs), DT (UTC -6hrs)[1]; Zona Noroeste Time: ST (UTC -8hrs, DT (UTC-7hrs).

[1]Some areas within this time zone do not change to daylight saving time (DT) in summer and remain on standard time (ST) all year-round.

Electrical voltage 127V (60Hz)

Public holidays Benito Juárez's birthday[1] (18 March 2024), Heroic Defense of Veracruz[2] (21 April 2024), Maunday Thursday[3] (28 March 2024), Good Friday[3] (29 March 2024).

[1]National holiday, [2]civic holiday, [3]religous festivities, not official holiday.

National tourist board website w visitmexico.com

Red tape A visa is not required for Australian, UK, EU or USA citizens to visit Mexico as a tourist for stays of up to 180 days. Other nationalities should check their visa requirements through their respective embassy/consulate (see w embassypages.com/mexico).

MOROCCO   *Telephone code +212*

Neighbouring countries Spain, Algeria, Western Sahara
Area 446,550 km^2 (excluding disputed territory of Western Sahara)
Climate The Atlantic coast experiences a mild climate with summer being the driest season. Along the Mediterranean coast winters are mild with rain at times. The interior of the country is hot in summer and mild in winter and drier than the coast.
Status Unitary parliamentary semi-constitutional monarchy
Population 36,767,655 (2022 estimate)
Life expectancy 74.3 years for women, 71.7 years for men (WHO, 2020–21)
Capital Rabat (population 577,827 / metropolitan area 2,120,192; 2014)
Other main towns and cities (population) Casablanca (3,359,818 / metropolitan area 4,270,750; 2014), Fez (1,112,072; 2014), Tangier (947,952; 2014), Marrakesh (928,850; 2014), Salé (890,403; 2014)
Economy (major industries) Imports: US$44,601 million (2019–20). Exports: US$36,578 million (2021) – agricultural produce, semi-processed goods (including textiles), phosphates and phosphate products. Data from the UN.
GDP US$3,896 per capita; US$142.874 billion (2022 estimate)
Languages Moroccan Arabic, Hassaniya Arabic, Berber, French
Religion Sunni Muslim (99.23%), Shia Muslim (0.45%), other (0.3%)
Currency Moroccan dirham (MAD)
Exchange rate £1 = 12.45MAD, US$1 = 10.35MAD, €1 = 11.06MAD (Feb 2023)

MOROCCO HIGHLIGHTS

- **Caves of Hercules** (Near Tangier) Cave complex, part natural and part manmade.
- **Cape Spartel** Northwestern-most point of Africa.
- **Meknes** Intimate imperial city – UNESCO World Heritage Site.
- **Tangier** Medina (old city).
- **Chefchaouen** Town known for its buildings painted different shades of blue. The narrow streets of the medina give a calm and peaceful atmosphere.

National airline Royal Air Maroc

Airports Casablanca-Mohammed V International, Marrakesh Menara, Agadir – Al Massira, Fes-Saïss, Tangier Ibn Battouta

Time UTC +1 hour, changing to UTC +0hrs during Ramadan month (start date is variable and follows the Islamic lunar calendar)

Electrical voltage 220V (50Hz)

Public holidays Feast of the Throne (30 July 2027), Anniversary of the Recovery Oued Ed-Dahab (14 August 2027), The Prophet Muhammad's Birthday (15 August 2027), The Prophet Muhammad's Birthday holiday (16 August 2027), The Anniversary of the Revolution of the King and the People (20 August 2027), Youth Day (21 August 2027)

National tourist board website w visitmorocco.com

Red tape A visa is not required for Australian, UK, EU or USA citizens to visit Morocco as a tourist for up to 3 months. Useful websites: w consulat.ma; w embassypages.com/morocco.

PORTUGAL   *Telephone code +351*

Neighbouring countries Spain

Area 92,212km²

Climate The climate is temperate with the north being cooler and sunshine more limited, compared with the warmer south.

Status Unitary semi-presidential republic

Population 10,352,042 (2021 estimate)

Life expectancy 84.4 years for women, 78.6 years for men (WHO, 2020–21)

Capital Lisbon (population 544,851 / metropolitan area 2,871,133; 2021)

Other main towns and cities (population) Porto (291,962 / metropolitan area 2,421,395; 2021), Vila Nova de Gaia (302,295; 2011), Amadora (171,719; 2011), Braga (193,333; 2021), Funchal (105,795; 2021)

Economy (major industries) Imports: US$77,515 million (2019–20). Exports: US$75,074 million (2021) – cars, motor vehicle parts, refined petroleum, leather footwear, packaged medication. Data from the UN.

GDP US$24,910 per capita; US$255.9 billion (2022)

Languages Portuguese

Religion Catholic (80.2%), no religion (14.1%), other Christian (4.6%), other (1.1%)

Currency Euro (EUR)

◀ *1 The Edge of the World is a natural landmark near Riyadh, Saudi Arabia. 2 Buildings in the 'blue city' of Chefchaouen, Morocco. 3 Colourful Ottoman-style houses in Berbera, Somaliland. 4 Amphitheatre at the ruins of Leptis Magna, Libya. 5 The Paiva Walkway by the Paiva River, Portugal. 6 Cenote Zaci, Valladolid, Mexico.*

- **Porto** Old town of Ribeira to newer grand plazas.
- **Braga** Said to be the most religious town in Portugal with many churches and holy monuments.
- **Guimaraes** Gothic buildings and medieval centre.
- **Paiva Walkway** An 8km-long wooded stairs and pathway with suspension bridges along the Paiva River.
- **Amarante** Picturesque old city set along Tamega River.

Exchange rate £1 = €1.13, US$1 = €0.94 (Feb 2023)
National airline TAP Air Portugal
Airports Faro, Lisbon, Porto
Time Winter UTC +0hrs (Western European Time, WET) from end of October to end of March. Summer, clocks move forward to UTC+1 hour (Western European Summer Time, WEST).
Electrical voltage 230V (50Hz)
Public holidays Assumption Day (15 August 2026), Our Lady of Sorrows[1] (20 August 2026), Our Lady of Graces[2] (22 August 2026).
[1]District holiday, Viana do Castelo; [2]District holiday, Braganca.
National tourist board website w visitportugal.com
Red tape A visa is not required for Australian, UK, EU or USA citizens to visit Portugal as a tourist for up to 90 days. Useful websites: w schengenvisainfo. com; w embassypages.com/portugal.

SAUDI ARABIA *Telephone code +966*

Neighbouring countries Jordan, Iraq, Yemen, Egypt, Kuwait, Qatar, United Arab Emirates, Oman

- **Riyadh** The cultural hub of Saudi Arabia – Al Masmak Fortress; Riyadh's National Museum.
- **Jeddah** Cosmopolitan city, rich in culture and historical attractions such as Jeddah's Floating Mosque.
- **Mecca** Great Mosque of Mecca, Islam's most important holy site.
- **Jabal al-Nour** The Cave, the sanctum of holy prophet Muhammad.
- **Khamis Mushait** The Green Mountain cable-car ride for stunning views of the city.

Area 2,149,690km^2

Climate Saudi Arabia has a desert climate and consequently rainfall is limited and mostly confined to the autumn/winter months. Conditions are very hot throughout the year.

Status Unitary Islamic absolute monarchy

Population 38,401,000 (2022 estimate)

Life expectancy 76.1 years for women, 73.1 years for men (WHO, 2020–21)

Capital Riyadh (population 7,676,654; 2018)

Other main towns and cities (population) Jeddah (4,697,000; 2021), Mecca (2,042,000; 2020 estimate), Medina (1,488,782; 2020), Dammam (1,252,523 / metropolitan area 4,140,000; 2020), Abha (1,093,705; 2021), Ha'il (936,465; 2021)

Economy (major industries) Imports: US$144,334 million (2019–20). Exports: US$267,546 million (2021) – crude petroleum, refined petroleum, ethylene polymers. Data from the UN.

GDP US$27,900 per capita; US$1.01 trillion (2022 estimate)

Languages Arabic

Religion Muslim 93.0%, Christian 4.4%, Hindu 1.1%, Buddhist 0.3%

Currency Saudi riyal (SAR)

Exchange rate £1 = 4.51SAR, US$1 = 3.75SAR, €1 = 4.01SAR (Feb 2023)

National airline Saudia

Airports King Abdulaziz (Jeddah), King Khalid (Riyadh), King Fahd (Dammam)

Time UTC +3hrs (Arabia Standard Time, AST)

Electrical voltage 220V (60Hz)

Public holidays None during the eclipse period

National tourist board website w visitsaudi.com

Red tape You will need a visa to visit Saudi Arabia if you are a citizen from Australia, UK, EU or USA. You can apply for a visa online or through your local Saudi Arabian embassy/consulate. Standard processing time is five days. Useful websites: w visitsaudi.com (online visas); w embassypages.com/saudiarabia.

SOMALIA   *Telephone code +252*

Neighbouring countries Ethiopia, Eritrea, Kenya

Area 637,657km^2

Climate The climate is desert or semi-desert with some very dry areas especially along the northern coast. There are two rainy seasons with very little rain falling outside these periods although more rain does fall in the northwest mountains. It is hot or very hot all year.

Status Federal parliamentary republic

- **Berbera** Daallo Escarpment with views over the Gulf of Aden – dense forest with some trees over 1,000 years old.
- **Xaafuun** (Hafan) Jutting out into Guardafui Channel on a sand spit it is the easternmost point of Africa and is home to numerous ancient structures and ruins.

Population 17,066,000 (2022 estimate)
Life expectancy 59.2 years for women, 54.0 years for men (WHO, 2020–21)
Capital Mogadishu (population 2,388,000; 2021)
Other main towns and cities (population) Hargeisa (1,200,000; 2019), Borama (398,609; 2014 estimate), Kismayo (234,852; 2014), Merca (230,100; 2014), Jamaame (129,149; 2005)
Economy (major industries) Imports: US$2,426 million (2019–20). Exports: US$164 million (2021) – gold, sheep and goats, insect resins, oily seeds. Data from the UN.
GDP US$544 per capita; US$5.218 billion (2022 estimate)
Languages Somali Arabic, English, Italian
Religion Muslim (99.8%), other (0.2%)
Currency Somali shilling (SOS)
Exchange rate £1 = 685SOS, US$1 = 569SOS, €1 = 608SOS (Feb 2023)
National airline Somali Airlines
Airports Mogadishu, Hargeisa
Time UTC +3hrs (East Africa Time, EAT)
Electrical voltage 220V (50Hz)
Public holidays The Prophet's Birthday (14–15 August 2027)
National tourist board website w somalia-tourism.gov
Red tape Visas are required to visit Somalia. The best time to apply for a visa is one to two months before your travel date. Useful websites: w visa.gov.so; w embassypages.com/somalia.

SPAIN   *Telephone code +34*

Neighbouring countries Portugal, France, Andorra
Area 505,990km^2
Climate Spain has three main climate types from the temperate and unsettled north, the arid and hot interior, during summer, and the Mediterranean conditions of the south and east coasts.
Status Unitary parliamentary constitutional monarchy
Population 47,163,418 (2022 estimate)

Life expectancy 85.7 years for women, 80.7 years for men (WHO, 2020–21)
Capital Madrid (population 3,223,334 / metropolitan area 6,791,667; 2018)
Other main towns and cities (city population/metropolitan area) Barcelona (1,636,762 / 5,474,482; 2019/2018), Valencia (794,288 / 2,522,383; 2019/2021), Seville (688,592 / 1,519,639; 2019/2021), Bilbao (1,037,847; 2018), Málaga (571,026; 2018), Zaragoza (675,301; 2021)
Economy (major industries) Imports: US$324,995 million (2019/20). Exports: US$391,558 million (2021) – cars, refined petroleum, packaged medication. Data from the UN.
GDP US$29,198 per capita; US$1.389 trillion (2022 estimate)
Language Spanish
Religion Christian (55.4%), no religion (40.3%), other (2.8%)
Currency Euro (EUR)
Exchange rate £1 = €1.13, US$1 = €0.94 (Feb 2023)
National airline Iberia Airlines
Airports Barcelona, Madrid, Málaga, Palma De Mallorca
Time UTC +1 hour (Central European Time, CET), from end of October to end of March. Clocks then change to Central European Summer Time (CEST) UTC +2hrs.
Electrical voltage 230V (50Hz)
Public holidays Feast of St James the Apostle[1] (25 July), Day of Institutions[2] (28 July), Feast of the White Virgin[3] (5 August), The Day of Our Lady of Africa[4] (5 August), Assumption of Mary (15 August), Assumption observed[5] (southern Spain: 16 August 2027).

[1]Autonomous community holiday in Basque Country, Galicia and Navarre; [2]Autonomous community holiday for Cantabria; [3]Provincial holiday in Álava; [4]Autonomous community holiday in city of Ceuta; [5]National holiday.

National tourist board website w spain.info

SPAIN HIGHLIGHTS

- **A Coruña** Known as the 'glass' city due to the unique architectural style of many of its buildings; Tower of Hercules lighthouse, dating to AD117.
- **Madrid** Plaza Mayor, historic and architectural landmark; Royal Palace of Madrid.
- **Alcázar Royal Palace** (Seville) Medieval Islamic palace, features in *Game of Thrones*.
- **Cádiz** Roman amphitheatre.
- **Tarifa** Boat trips to observe dolphins and fin, orca, pilot and sperm whales.

Red tape Visas are not required for Australian, UK, EU or USA citizens to visit Spain as a tourist for up to 90 days. Useful websites: w schengenvisainfo. com; w embassypages.com/spain.

TUNISIA   *Telephone code +216*

Neighbouring countries Algeria, Libya

Area 163,610km²

Climate Tunisia's climate is Mediterranean along its northern and eastern coasts with hot summers and mild winters, while the interior is desert or semi-desert with very hot summers and mild winters.

Status Unitary presidential republic

Population 11,708,370 (2020 estimate)

Life expectancy 79.2 years for women, 74.9 years for men (WHO, 2020–21)

Capital Tunis (population 602,560 / metropolitan area 2,658,816; 2022)

Other main cities/towns (population/metropolitan area) Sfax (330,440 / 1,018,341; 2014), Sousse (271,428 / 674,971; 2013), Ettadhamen-Mnihla (196,298; 2014), Kairouan (187,000; 2014), Gabès (152,921; 2014)

Economy (major industries) Imports: US$18,446 million (2019–20). Exports: US$16,432 million (2021) – insulated wire, pure olive oil, non-knit men's/women's suits. Data from the UN.

GDP US$3,763 per capita; US$45.642 billion (2022 estimate)

Languages Tunisian Arabic, Berber, French, Judeo-Tunisian

Religion Muslim (99%), other including Christian and Jewish (1%)

TUNISIA HIGHLIGHTS

- **El Djem** Roman amphitheatre set against surrounding town.
- **Carthage** Phoenician ruins in suburbs of Tunis.
- **Sidi Bou Said** Beautiful clifftop village in suburbs of Tunis, popular with artists.
- **Grand Erg Oriental** (Nr Douz) Magnificent sand dunes.
- **Chott el Djerid** (Nr Tozeur, 211km west of Gabès) 'Moonscape' desert in dry season.
- **Ribat of Monastir** (23km southwest of Sousse) 8th-century fortress built during the Abbasid conquest. Better known in more recent times as the location of the 1979 Monty Python film *Life of Brian*.

1 Dragon blood trees on Socotra Island, Yemen. 2 Tarifa on Spain's southern coast is a great location for whale-watching. 3 A memorial marking the site of the battle of the Alamo, Texas, USA. 4 Monastir's medieval fortress, Tunisia. ▶

OLEG ZNAMENSKIY/S

MILVUS80/S

SALVADOR AZNAR/S

ROMAS_PHOTO/S

Currency Tunisian dinar (TND)

Exchange rate £1 = 3.78TND, US$1 = 3.14TND, €1 = 3.35TND (Feb 2023)

National airline Tunisair

Airport Tunis-Carthage

Time UTC +1 hour (CET)

Electrical voltage 230V (50Hz)

Public holidays Republic Day (25 July 2027), The Prophet's Birthday (14 August 2027)

National tourist board website w discovertunisia.com

Red tape Visas are not required for Australian, UK, EU or USA citizens to visit Tunisia as a tourist for up to 90 days. Application will need to be through your local embassy/consulate if a visa is required for longer (see w embassypages.com/tunisia).

USA *Telephone code +1*

Neighbouring countries Canada, Mexico

Area 9,833,520km^2

Climate The vast size of the USA means there are a number of climate types across the country but in general the climate is continental with hot summers and cold winters. However, in areas near the Atlantic/Pacific oceans or on plateaus and mountainous regions the climate is tempered and less extreme.

Status Federal presidential constitutional republic

Population 331,449,281 (2020)

Life expectancy 80.7 years for women, 76.3 years for men (WHO, 2020–21)

Capital Washington, DC (population 689,545 / metropolitan area 6,385,162; 2020)

Other main cities/towns (population/metropolitan area) New York City (8,804,190 / 20,140,470; 2020), Los Angeles (3,898,747 / 13,200,998; 2020), Chicago (2,746,388 / 9,618,502; 2020), Houston (2,304,580 / 7,122,240; 2020), Phoenix (1,608,139 / 4,845,832; 2020)

Economy (major industries) Imports: US$2,407,543 million (2019–20). Exports: US$1,753,941 million (2021) – refined/crude petroleum, cars, integrated circuits, vehicle parts. Data from the UN.

GDP US$75,180 per capita; US$25.35 trillion (2022)

Languages English (de facto)

Religion Protestant (40%), no religion (29%), Catholic (21%), other Christian (2%), other (6%)

Currency US dollar (US$)

Exchange rate £1 = US$1.20, €1 = US$1.07 (Feb 2023)

Airlines American Airlines, Delta Air Lines, United Airlines, Air Canada (note: there is no national airline)

- **San Antonio** (Texas) The Alamo (site) museum of the famous battle that was the crossroads in Texan history.
- **Dallas** (Texas) Metropolitan City; Frontiers of Flight Museum; Perot Museum of Nature and Science.
- **Hot Springs** (Arkansas) Hot Springs National Park in an urban setting; Ouachita National Forest.
- **New York City** Summit the Empire State Building; visit the American Museum of Natural History; ferry to see the Statue of Liberty.

Airports Atlanta, Los Angeles, Chicago, Dallas

Time Eastern Time: EST (UTC -5hrs), EDT (UTC -4hrs); Central Time: CST (UTC -6hrs), CDT (UTC -5hrs); Mountain Time: MST (UTC -7hrs), MDT (UTC -6hrs)[1]; Pacific Time: PST (UTC -8hrs), PDT (UTC -7hrs). [1]Some regions within this time zone remain on MST all year round, eg: most of Arizona. All time zones standard time (ST) run from November to March and daylight-saving time (DT) from March to November. List of US time zones presented is not exhaustive.

Electrical voltage 120V (60Hz)

Public holidays None during the eclipse period

National tourist board website w visittheusa.co.uk

Red tape The US Visa Waiver Program (VWP) enables citizens from the UK, Australia and EU countries to travel to the USA for stays of up to 90 days as tourists, providing certain requirements are met. On average the standard processing time is seven to ten working days. Useful websites:w visittheusa.co.uk; w embassypages.com/usa.

YEMEN   *Telephone code +967*

Neighbouring countries Saudi Arabia, Oman

Area 555,000km^2

Climate The climate is hot and dry with only sporadic rainfall in low lying and coastal regions. On the western plateau conditions are tempered with higher rainfall.

Status Unitary provisional government

Population 30,984,689 (2022 estimate)

Life expectancy 68.9 years for women, 64.4 years for men (WHO, 2020–21)

Capital Sana'a (population 2,545,000; 2017)

Other main cities/towns (population) Ta'izz (2,612,000; 2014), Al Hudaydah (402,560; 2004), Aden (863,000; 2017), Ibb (160,000; 2005), Dhamar (146,346; 2004)

- **Sana'a** Old city, with a wealth of intact ancient archaeology – UNESCO World Heritage Site.
- **Aden** Tower of Silence: ruins of a Zoroastrian Temple once used for excarnation burials.
- **Jiblah** Queen Arwa Mosque, one of the oldest in Yemen and last resting place of Queen Arwa.
- **Socotra Island** Hoq Cave; Qalansiyah beach and lagoon.

Economy (major industries) Imports: US$8,995 million (2019–20). Exports: US$1,800 million (2021) – crude petroleum, gold, non-fillet frozen fish, industry fatty acids. Data from the UN.

GDP US$891 per capita; US$28.134 billion (2022 estimate)

Languages Arabic

Religion Muslim (99%), other (1%)

Currency Yemeni rial (YER)

Exchange rate £1 = 301YER, US$1 = 250YER, €1 = 267YER (Feb 2023)

National airline Yemenia

Airports Sana'a, Aden, Seiyun

Time UTC +3hrs (AST)

Electrical voltage 230V (50Hz)

Public holidays None during the eclipse period

National tourist board website w yementourism.com

Red tape Visas are required for all visitors to Yemen and are required prior to travelling. Application should be through your local embassy/consulate. On average the processing time is two weeks. Useful websites: w yemenembassynl.org; w embassypages.com/yemen.

INDEX OF ADVERTISERS

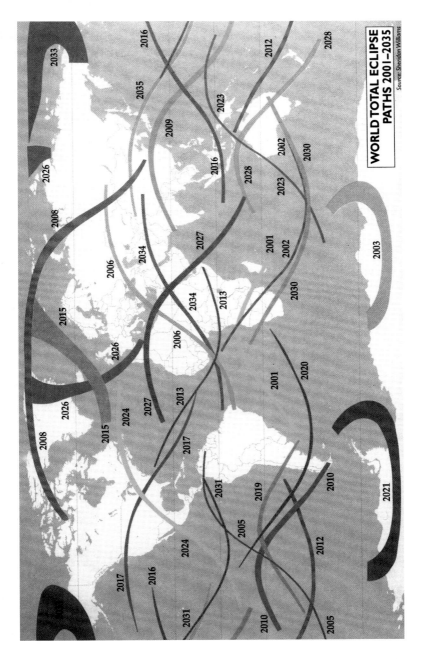

WORLD TOTAL ECLIPSE PATHS 2001–2035

Source: Sheridan Williams

127

THE BRADT STORY

In the beginning

It all began in 1974 on an Amazon river barge. During an 18-month trip through South America, two adventurous young backpackers – Hilary Bradt and her then husband, George – decided to write about the hiking trails they had discovered through the Andes. *Backpacking Along Ancient Ways in Peru and Bolivia* included the very first descriptions of the Inca Trail. It was the start of a colourful journey to becoming one of the best-loved travel publishers in the world; you can read the full story on our website (bradtguides.com/ourstory).

Getting there first

Hilary quickly gained a reputation for being a true travel pioneer, and in the 1980s she started to focus on guides to places overlooked by other publishers. The Bradt Guides list became a roll call of guidebook 'firsts'. We published the first guide to Madagascar, followed by Mauritius, Czechoslovakia and Vietnam. The 1990s saw the beginning of our extensive coverage of Africa: Tanzania, Uganda, South Africa, and Eritrea. Later, post-conflict guides became a feature: Rwanda, Mozambique, Angola, and Sierra Leone, as well as the first standalone guides to the Baltic States following the fall of the Iron Curtain, and the first post-war guides to Bosnia, Kosovo and Albania.

Comprehensive – and with a conscience

Today, we are the world's largest independently owned travel publisher, with more than 200 titles. However, our ethos remains unchanged. Hilary is still keenly involved, and **we still get there first**: two-thirds of Bradt guides have no direct competition.

But we don't just get there first. Our guides are also known for being **more comprehensive** than any other series. We avoid templates and tick-lists. Each guide is a one-of-a-kind expression of an expert author's interests, knowledge and enthusiasm for telling it how it really is.

And a commitment to wildlife, conservation and respect for local communities has always been at the heart of our books. Bradt Guides was **championing sustainable travel** before any other guidebook publisher. We even have a series dedicated to Slow Travel in the UK, award-winning books that explore the country with a passion and depth you'll find nowhere else.

Thank you!

We can only do what we do because of the support of readers like you – people who value less-obvious experiences, less-visited places and a more thoughtful approach to travel. Those who, like us, take travel seriously.

Bradt GUIDES

TRAVEL TAKEN SERIOUSLY